AUSTRALIAN TECHNOLOG

TEXTILES AND TECHNOLOGY

KATE BAULCH
KRIS OPPERMANN

Series Editor: Ray Stevens

CAMBRIDGE
UNIVERSITY PRESS

Published by the Press Syndicate of the University of Cambridge
The Pitt Building, Trumpington Street, Cambridge CB2 1RP, UK
40 West 20th Street, New York, NY 10011-4211, USA
10 Stamford Road, Oakleigh, Melbourne 3166, Australia

First published 1994

Printed in Australia by Australian Print Group

National Library of Australia cataloguing in publication data
Baulch, Kate
Textiles and technology.
Includes index.
1. Textile crafts. 2. Textile crafts — Problems, exercises, etc.
3. Technology. 4. Technology — Problems, exercises, etc. I.
Oppermann, Kris. II. Title.
746

Library of Congress cataloguing in publication data
Baulch, Kate.
Textiles and technology/Kate Baulch, Kris Oppermann.
p. cm. — (Australian technology studies).
Includes index.
1. Textile fabrics. I. Oppermann, Kris, 1967– . II. Title.
III. Series.
TS1455.B35 1994
677—dc20 94-3837
CIP

A catalogue record for this book is available from the British Library.

ISBN 0 521 43784 9 Paperback

Central Cover Photograph:
Anton Geisser/Scoopix

ACKNOWLEDGEMENTS

The authors would like to thank Julie Sultana for her student case study and Dianna Walker for assistance with computer technology and ongoing support. We would also like to thank Lauretta Pozzuto, Marie Haslam, Michelle Lewis and the other Technology Studies students for their assistance at different times, our friends and families for their assistance and support, and our colleagues at Marian College, West Sunshine.

Grateful acknowledgement is made to the following for their assistance in both providing suitable material and granting Cambridge University Press permission to reproduce it in this text.

Photograph (Fig. 1.4), D W Hamilton, The Image Bank; illustration (Fig. 1.6), Australian Cotton Foundation Limited; photograph (Fig. 2.4) Pfaff Australia; illustration (Fig. 3.1), Reserve Bank of Australia; illustration (Fig. 3.5) International Wool Secretariat — the Woolmark, the symbol of pure new wool products, is a trade mark owned by IWS and may only be used by licensed manufacturers on products meeting appropriate performance specifications; illustration (Fig. 3.6), IWS — the Woolblendmark, the symbol of new wool rich products, is a trade mark owned by IWS and may only be used by licensed manufacturers on products meeting appropriate performance specifications; photograph (Fig. 3.8), Australian Wool Corporation; photographs (Fig. 3.9 & 3.10), CSIRO, Division of Wool Technology; illustration (Fig. 3.11), AWC & CSIRO; illustration (Fig. 5.2), Standards Australia Quality Assurance Services; newspaper article (Fig. 5.5), Janice Breen Burns, The Age; text and illustrations (Fig. 5.6), Paddy Pallin; photograph (Fig. 6.11), Linda Jackson; photograph (Fig. 8.2), International Photographic Library; photograph (Fig. 8.4), Image Solutions; photograph (Fig. 8.3), David Jones; photograph (Fig. 8.10), David Tatnall.

Every effort has been made to trace and acknowledge copyright but in some cases this has not been possible. Cambridge University Press would welcome any information that would redress this situation.

CONTENTS

NOTES FOR TEACHERS

This book is aimed primarily at teachers and students of textiles but has many links with other areas of design and technology, particularly in Designing and Investigations. Teachers will find it useful for:

- classroom activities
- planning courses
- reference material

It is also a student reference and activity book. Each chapter is accompanied by realistic student activities which will challenge and extend every student's learning and understanding of Design and Technology.

Sections of the book have clear links with VCE Technology requirements (e.g. Materials and Technology: testing and evaluation, Technological Design and Development: design brief, production and evaluation). However, it has been written so that it can be used in any Textiles or Technology Studies program.

The book is highly illustrated. The processes it describes are clearly set out to provide students with an easy to follow and ready to use format. The focus is on problem solving by design and most of the student activities reflect this approach to learning.

The book belongs to the Australian Technology Studies series. You will find many links between this and the other books in the series:

- Materials, Design and Technology
- Wood and Technology
- Plastics and Technology
- Metals and Technology

AN INTRODUCTION TO TEXTILES

We are constantly undergoing vast technological changes in the area of textiles. Since the advent of synthetic fibres we have seen many changes in materials, tools, equipment and therefore in the use of textiles. For thousands of years, people have been clothing themselves for protection, adornment and decoration. Firstly with animal skins, later with twisted and interlaced leaves and stems of plants. Obviously these items didn't fit very well, let alone provide warmth or comfort. But they fulfilled a requirement. Humans then learnt to make fibres from these raw materials and turn them into cloth. Since this time there have been huge changes in technology, and in textiles—both internationally and within Australia. This chapter will look at some of those changes and developments.

This chapter will focus on:

- what are textiles?
- what are fibres?
- from fibre to fabric—a brief look at some processes.

WHAT ARE TEXTILES?

The word 'textile' comes from the Latin word 'texere', which means 'to weave'. Theoretically, textile still means this. However, we now apply it in much broader terms to all fibres, yarns and other materials which are made into fabrics. Some key terms to keep in mind are:

- FIBRES, the raw materials used in textiles
- YARN, what the fibres are made into
- FABRICS, the materials which are made from the yarn.

All of these things are included in the term 'textiles'.

Textiles in various forms have been evident in the world for thousands of years. The fibres used and methods of production have changed dramatically over the past century—mainly through industrialisation and the development of synthetic fibres. Due to advances in technology, we now have a range of textiles with many domestic and industrial uses.

Clothing is perhaps the most widely recognised form of textiles today. However, textiles are also evident in many other places: households, as furnishings, curtains, cushions and carpets; industry, making tarpaulins and linings for example; and in leisure goods, such as sails for boats, balloons and backpacks. We cannot avoid coming into contact with products made from textiles every day of our lives.

CLOTHING

Clothing has many functions: it keeps us warm, provides protection and adds comfort. It can also tell us something about people's personalities and even their jobs. We select our clothing for many reasons, such as:

- comfort
- status
- decoration
- cost
- expressing individuality.

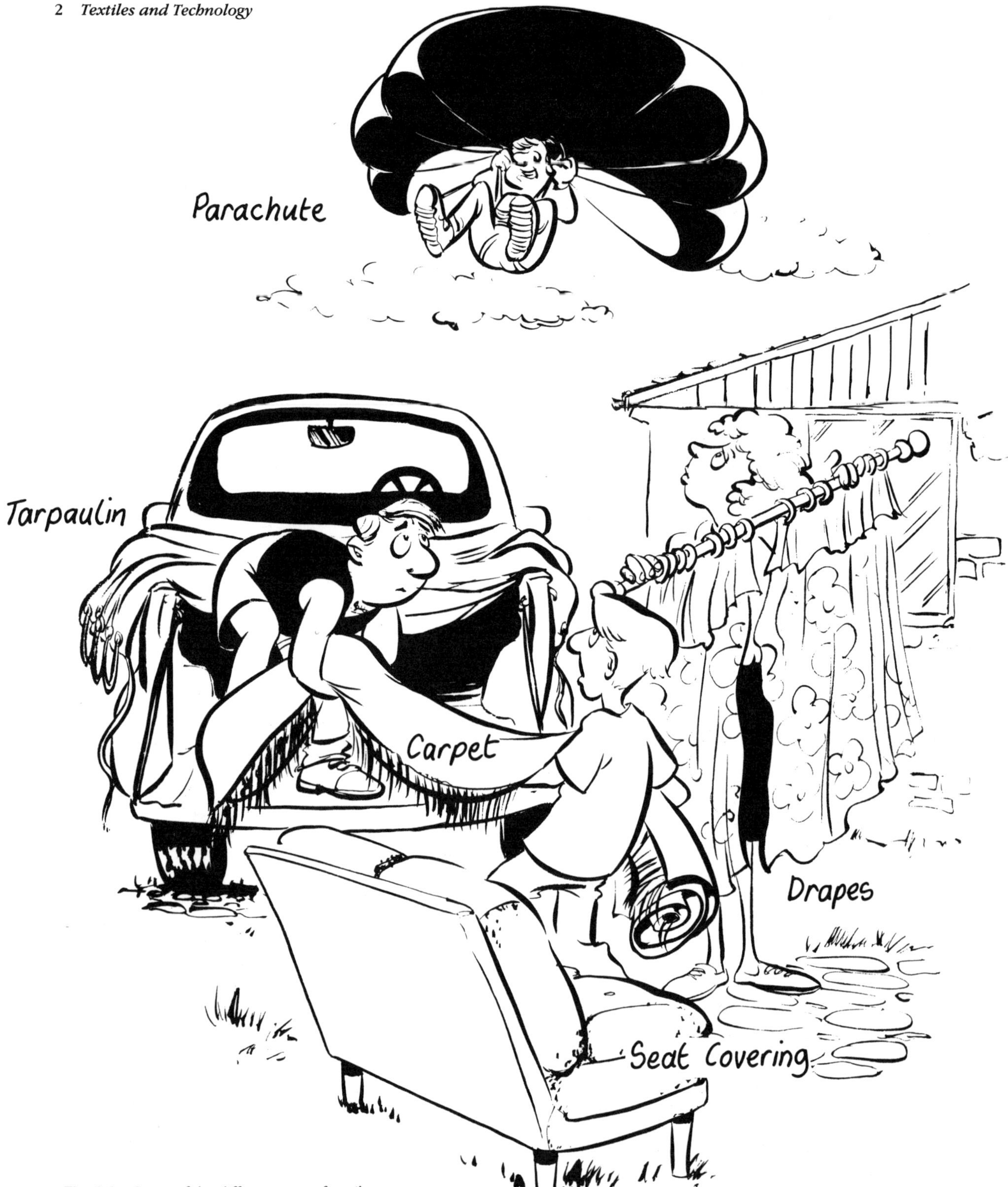

Fig. 1.1 Some of the different uses of textiles

We change our type of clothing dramatically over our lifetimes—partly due to fashion but also as our personalities develop. Consider the style of clothing you wear now. Is it the same as what you were wearing two years ago? What has changed? Would you wear the clothes your mother or older sister does?

Clothing changes as new fabrics and processes become available. In the past, only the rich were able to be fashionable. Now, due to cheaper fabrics, more people are able to keep up with fashions if they wish to. The introduction of synthetic fibres has led to vast changes in the properties of fabrics (consider all the drip dry fabrics now on the market). There are many different styles of clothing now. And nearly anything is acceptable!

Fashion is very competitive—many would argue it has become too competitive. A huge international industry has developed. There are large numbers of people associated with the industry. In Australia alone—think of the shopkeepers, fashion designers, textile designers and factory workers. We will now have a brief look at fibres and some of the ways these fibres are processed to form fabrics.

An English Lady, circa 1770 A French Nobleman, circa 1770 Clothing in the early twentieth century, circa 1916

Fig. 1.2 The development of clothing

Fig. 1.3 Different styles of clothing in the 1990s

FIBRES

One of the main forms of textiles are fabrics. Fabrics are made from fibres, which are processed and then usually either knitted or woven to form the fabrics. Historically, fibres have been natural—coming from plants or animals. Nowadays we also have many mineral and synthetic fibres.

NATURAL FIBRES

Natural fibres come from three sources: plants, animals and minerals. We will briefly look at each. You may then wish to look at one or more of them in more detail. Table 1.1 lists some of the better known animal and plant fibres. Later in the chapter there is a list of some of these fibres' characteristics.

Table 1.1 Plant and Animal Fibres

Plant Fibres	Animal Fibres
cotton	wool
linen	silk
jute	mohair
hemp	angora
abaca	cashmere
sisal	alpaca
kapok	leather
	suede

PLANT FIBRES

Plant fibres are made of a woody substance called cellulose and are therefore referred to as cellulosic fibres. The best known are flax and cotton.

FLAX

Flax is a small plant with blue flowers. The leaves and stalks of the flax plant are made into linen, a strong and cool material used for making clothing, tablecloths, napkins and laces. Flax was the first plant to be grown for spinning and weaving, and probably the first to be used for making textiles as traces of it have been found from over 4000 years ago. To make linen, the plant needs to be pulled from the ground as some of the most valuable parts of the plant are in the lower stem and roots. The stalks are then soaked in water until the green part rots away. This process is called retting. The fibres that are left after retting can then be separated and dried. Once dry, they are scraped to remove any dirt, and combed to leave the fibres straight and parallel to one another—ready to be spun and woven into linen.

COTTON

The cotton plant grows to about 1.2–1.5 m in height. After flowering, a pod or 'boll' about the size of a walnut remains on the flowering stalk. When it is ripe, the boll bursts open and forms a ball of cotton wool, called a lint. The lint is picked and the cotton in the lint is separated from the hairs by beating it with a stick. This process is called ginning. The raw cotton is then packed tightly in bales and sent to mills.

Fig. 1.4 Flax

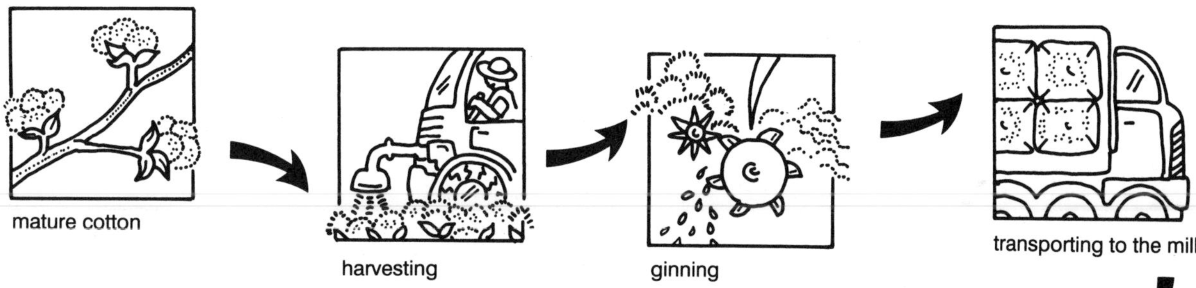

Once the cotton reaches the mills, the process of making it into yarn begins. Firstly, the fibres are graded into similar lengths and colours. The bales are then opened and the compressed fibres dusted and blended to help the next process: carding, combing or condensing. All three processes untangle the fibres. Once this is completed, drawing or drafting takes place to reduce the thickness of the fibres so they can be spun into yarn. You then have either combed cotton yarn, carded cotton yarn or condensed spun cotton yarn—depending on the process followed.

Cotton is now the most widely used textile fibre. It readily absorbs moisture and yet allows moisture to evaporate freely. It is also strong and hard wearing.

Cotton was first cultivated in India. It wasn't until early in the twentieth century that cotton was successfully grown in Australia. Prior to then, there were problems with climate and pests. Cotton was first grown commercially in Queensland, but really developed during the 1960s in NSW. Most of the cotton today is still grown in NSW, mainly around the Namoi and Macquarie regions. Due to increased investment, improved irrigation and technology, the industry now meets all our domestic needs for raw cotton and we export it, mainly to Japan, Taiwan and South Korea.

COTTON AUSTRALIA

Fig. 1.6 The 'Cottonmark' symbol

ANIMAL FIBRES

All animal fibres are made of protein and are therefore called proteinic fibres. The two best known animal fibres are wool and silk.

WOOL

Wool usually comes from sheep but can also come from other animals. It is a particularly important fibre for Australia as we are one of the greatest wool producing countries in the world.

The sheep are shorn—removing the wool in a single piece, called a fleece. There are many different types of wool, depending on the type of sheep. The Merino is

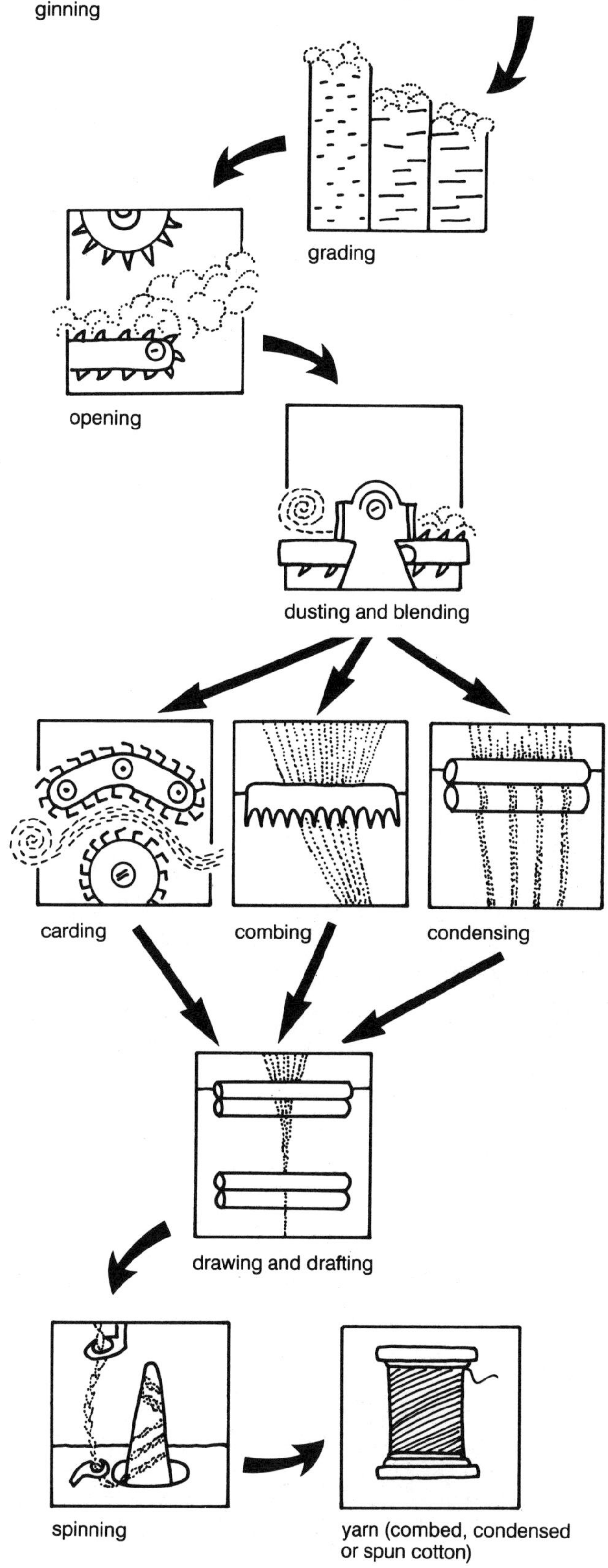

Fig. 1.5 Cotton: the process from cotton plant to yarn

the best known and is said to have the finest wool. One sheep will give different hairs: they can be long, short, straight or crimped (curly). The wool on a sheep's back is the coarsest as it protects the sheep from the weather. Chapter 3 deals in detail with the process from sheep farming to fabric.

Goats, camels, llamas and rabbits also provide hair or wool that can be spun and made into textiles. Angora is a wool made from rabbit fur and sheep's wool. Mohair comes from the angora goat and alpaca is wool from llamas.

SILK

Silk is produced by silkworms, the caterpillars of several species of moth that live in Mulberry trees. The silk comes from the caterpillar's cocoon. After hatching, silkworms feed for several weeks before spinning their cocoons. A liquid is secreted through a tiny hole, called a spinneret, and forms two fine filaments which are gummed together into a delicate thread. The silkworm then spins the thread around itself until it is eventually fully enclosed. At this stage, under natural conditions, the silkworm would finish the process and turn into a moth. In silk farms, they are killed and the cocoon is dropped into hot water to loosen the gum and give us access to the thread.

Silk is an expensive fabric as the process of getting the fibre and making it into thread is quite arduous. Silk is lustrous, shiny and strong. Japan and China are the most important silk producing countries.

Fig. 1.7 Silkworm feeding on mulberry leaves and its cocoon, from which silk is obtained

MINERAL FIBRES

This small group of fibres have fairly limited uses, but should still be noted. The two which are probably most common are aluminium and asbestos.

ALUMINIUM

Aluminium fibre is used to add glitter and sparkle to other fibres. These glittery fibres have in the past mainly been used for evening wear. Today we use a greater variety of fabrics and decoration techniques in casual wear and this can involve the use of shiny aluminium fibres.

ASBESTOS

Asbestos is also classed as a textile fibre although its properties are quite different. As asbestos is difficult to spin it is often combined with cotton. Because of its great fireproof qualities, asbestos was used in things which required insulation, such as oven cloths, fireproof clothing and ironing board covers. Recently, it was found that asbestos can cause asbestosis—a disease affecting the lungs—if someone is exposed to asbestos dust for extended periods. It has therefore had very limited use in recent times.

SYNTHETIC FIBRES

We have looked at natural fibres, but there are also those which are made either partly or completely from chemicals. These are divided into two categories:

- fibres made from a natural base (cellulose), such as acetate, viscose rayon and polyester
- fibres made from a chemical base, such as nylon and acrylic.

We will take a brief look at these two types of synthetic fibres and the development of some of those that are better known.

Table 1.2 Synthetic Fibres

Cellulosic	Non-Cellulosic
acetate	acrylic
triacetate	modacrylic
cuprammonium rayon	polyamide (nylon)
viscose rayon	nytril
polynosic	olefin
alginate	polyester (dacron, terylene)
	spardex (elastomeric, e.g. lycra, vyrene)
	polyvinyl alcohol (vinylon)
	polyvinyl chloride (PVC)

CELLULOSIC FIBRES

Cellulose is a chemical derived from plants. To make fibres for textiles, the cellulose is extracted and mixed with a number of chemicals. These chemicals vary, depending on the fibre to be made, but the end result is similar—a thick liquid which is then forced through tiny holes in a spinneret to form threads. These threads are then either hardened in acid and twisted into yarn or spun then twisted into yarn, depending on the type of fibre.

NON-CELLULOSIC FIBRES

These fibres are made from chemicals only. Tiny molecules are joined into units which are in turn built into long chains called polymers. This process is called polymerisation. A liquid is then formed by either heating or dissolving the chemicals, depending on the fibre being made. The liquid is forced through a spinneret and spun.

PROPERTIES OF FIBRES

Cotton

Fibre: fine, short staple fibre

Fabric: medium to hard, strong, cool fabric; poor wrinkle recovery; absorbent

Thermal Properties: conducts heat well; burns

Care: can be bleached; fair to good resistance to sunlight; attacked by micro-organisms; dry-cleanable, launder in cold to boiling water; withstands hot ironing

General: has good durability, is generally comfortable, low cost

Wool/Mohair

Fibre: fine to course, short to long staple fibre

Fabric: soft to medium, weak fabric; good wrinkle recovery; good to very good absorbency

Thermal Properties: poor conductor of heat; insulates well; smoulders

Care: avoid bleach; has poor resistance to sunlight (fades); attacked by micro-organisms and insects; can be dry-cleaned; launder in cold to hand warm water; warm iron

General: fair to good durability; comfortable; moderate cost

Flax (Linen)

Fibre: medium to course, medium to very long staple fibres

Fabric: medium to hard, crisp fabric; very strong; poor wrinkle recovery; good to very good absorbency

Thermal Properties: very good conductor of heat; burns

Care: can be bleached with care; fair to good resistance to sunlight; attacked by micro-organisms; can be dry-cleaned; launder in cold or hot water; hot iron

General: fair to good durability; comfortable; expensive

Silk

Fibre: fine filament

Fabric: medium weight, strong, warm fabric; fair wrinkle recovery; absorbent to very absorbent

Thermal Properties: poor conductor of heat; smoulders

Care: avoid bleach; poor resistance to sunlight; attacked by micro-organisms and insects; can be dry-cleaned; launder in cold to warm water; use a cool iron

General: fair durability; comfortable to very comfortable; very expensive

Nylon

Fibre: fine to course (as required) filament or staple fibre

Fabric: medium to hard, strong to very strong fabric; good to very good wrinkle recovery; poorly absorbent

Thermal Properties: poor conductor of heat; thermoplastic; burns or melts

Care: avoid bleach; fair resistance to sunlight (yellows); not attacked by micro-organisms or insects; can be dry-cleaned; launder in cold to warm water (yellows in hot water)

General: good to very good durability; fair comfortableness, but if body temperature increases slightly, the fabric may feel hot and clammy, should be worn with more absorbent fibres to enhance comfort; moderate cost

Polyester

Fibre: fine to course (as required) filament or staple fibre

Fabric: medium to soft fabric; fair strength; good to very good wrinkle recovery; very poor absorbency

Thermal Properties: poor conductor of heat; very thermoplastic; burns or melts

Care: resistant to sunlight; not attacked by micro-organisms or insects; launder in cold or hot water

General: good to very good durability; fair comfortableness, but if body temperature increases slightly, the fabric may feel hot and clammy, should be worn with more absorbent fibres to enhance comfort; moderate cost

Viscose

Fibre: fine to course (as required) filament or staple fibre

Fabric: soft, weak fabric; very poor wrinkle recovery; very good absorbency

Thermal Properties: poor conductor of heat; burns

Care: avoid bleach; poor to fair resistance to sunlight; attacked by micro-organisms; can be dry-cleaned; launder in cold or hot water (yellows)

General: poor durability; good comfortableness; low cost

Acetate

Fibre: fine filament or staple fibre

Fabric: fine fabric; poor wrinkle recovery; fair absorbency

Thermal Properties: poor conductor of heat; melts

Care: fair to good resistance to sunlight; not attacked by micro-organisms or insects; can be dry-cleaned; launder in cold to warm water (yellows in hot water)

General: poor durability; fair comfortableness; low to moderate cost

RAYON

In 1884, Count Hildaire de Chardonnet started to wonder what would happen if all the silkworms died. This led him to develop a replacement for silk. In 1889, he opened the first synthetic textiles factory. He called his fibre 'artificial silk' but it is better known today as rayon. Rayon was the first synthetic fibre, although there had been developments prior to its creation. Rayon has been used to make tassels, braids and sweaters.

Rayon is made from cellulose. The fibres come from wood which is sawn into very fine pieces and mixed with water to form wood pulp. The pulp is then soaked in caustic soda to untangle the extremely fine fibres. Carbon disulphide is added and changes the substance into solid yellow crumbs. This process is called xanthation. These 'crumbs' are soaked for several days in caustic soda. Provided the temperature and humidity are right, the fibres change into viscose. The viscose is then squirted into a bath of sulphuric acid, which changes it into pure cellulose fibres. The cellulose fibres can then be made into yarn.

NYLON

In the 1930s, scientists began to look for ways to make polymer fibres without using any wood or other natural substances. Two companies individually came up with a solution. In 1938, a team of scientists led by Dr Wallace Carothers at Du Pont Chemical Company in the USA came up with 'Fibre 6.6' later known as nylon. The Du Pont company began to produce nylon commercially. In 1940, twelve years after the research began, nylon stockings were sold to the public; the first nylon garments.

POLYESTER

In 1941, J.R. Quinfield and J.T. Dickson came up with a strong, synthetic fibre which could withstand very high temperatures. They called it polyester. The large European firm Imperial Chemical Industries (ICI) produced the fibre. Polyester has proved to be the most economically important synthetic fibre. Many new synthetic fibres have been developed, and are still being developed, from polyester. It is estimated that several new fibres reach world markets every week.

GLASS FIBRES

We have made things from glass for many years, and now use glass fibres in many different ways. Glass and glass fibres are not suitable to use for clothing but can be used for other types of textiles. Fabrics made from glass fibres have some very useful qualities: they are fireproof, dirt resistant, drape well and are easy to clean. Glass fibre materials are good for insulation against temperature and sound.

Glass fibres can be bonded with plastic to make fibreglass. Fibreglass can be used for a variety of things such as boats, skis and surfboards.

FROM FIBRE TO FABRIC

Fibres are made into fabrics in a number of ways. We will have a brief look at how and some of the machinery and equipment used. Later, this chapter looks at the materials, processes and equipment in more detail. Chapter 3 looks at the processing of wool and will give you more details of these processes.

CARDING

Carding is brushing the fibres with a wire brush. It untangles the fibres so that they can be spun.

SPINNING

Once we have untangled the fibre, it needs to be made into a long continuous thread or yarn. This is usually done by spinning (twisting) the fibres together.

WEAVING

An examination of either your wardrobe or home furnishings will quickly reveal that most fabrics are either woven or knitted, with the exception of felt, lace and net. Woven fabrics are made on looms by weaving or interlacing two sets of yarns at right angles to each other: lengthwise, the warp and crosswise, the weft.

There are many different types of weaves based on three things:

- the thickness of the yarn
- length over and under
- the type of loom used.

Weaving was originally done on small hand looms. Nowadays, we have much larger and more efficient looms which allow us to make more varied fabrics at a much faster rate.

There are three fundamental weaves on which all others are based: plain, twill, and satin. It is important that you become familiar with the many kinds of materials available. Aim to collect as many samples of each as possible.

PLAIN WEAVE

Lawn, gingham, organdie, chintz and handkerchief linen are examples of plain weave. These fabrics are generally strong and durable due to their firm construction as they are closely woven to produce maximum interlacing. There is no right or wrong side unless printed or specially finished.

Possible variations are Basket Weave and Rib Weave.

TWILL WEAVE

Cotton drill, denim, flannelette, woollens, worsted, gaberdine, serge and some suiting fabrics are examples of twill weave. Twill always shows diagonal ridges across the fabric which can run from left to right or, like most cotton twills, right to left. Twills have increased bulk, warmth and strength because of the

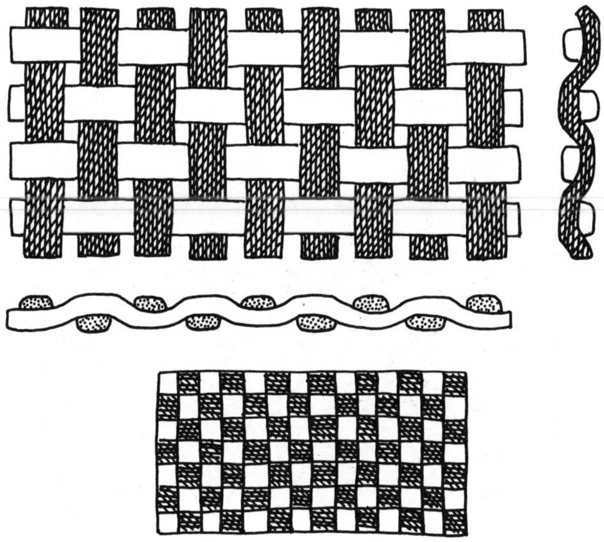

Fig. 1.8 Plain Weave

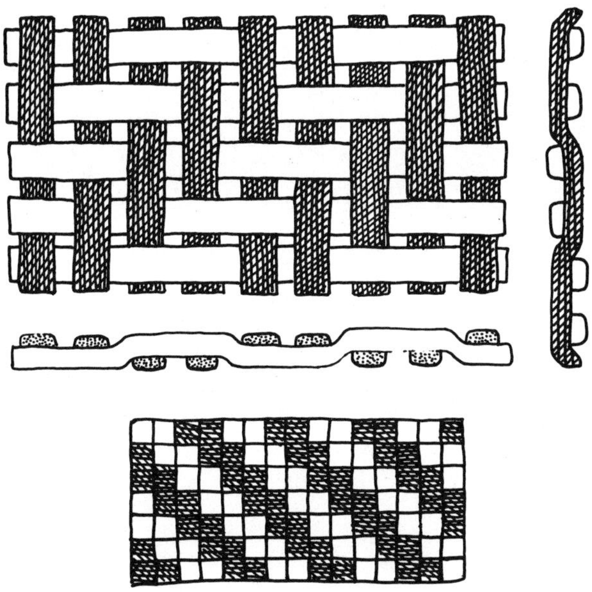

Fig. 1.9 Twill Weave

closeness of the warp threads. This makes them great for work and sport. Twill fabrics have a right side and a wrong side.

Variations of twill weave are Herringbone and Birdseye.

SATIN

Satin weave is quite irregular and characterised by long, floating warp yarns lying on the surface of the right side of the fabric. These threads are called 'floats'. Obviously the longer the floats the less durable the material as they tend to catch or snag on rough surfaces. It is these long floating threads that catch the light and give satin its characteristic shiny appearance. Silk and rayon are the most common fibres used for this process.

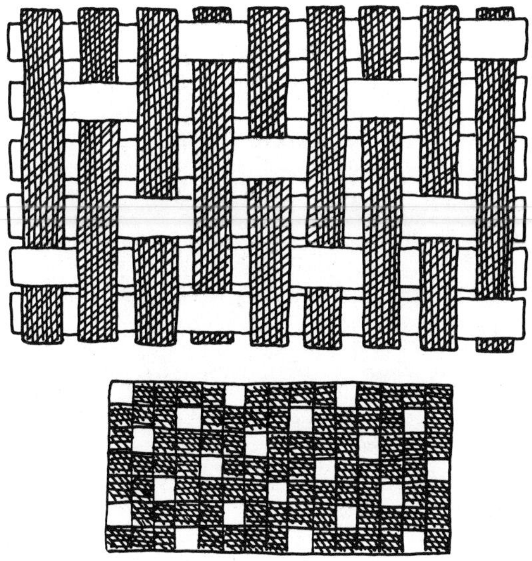

Fig. 1.10 Satin Weave

OTHER WOVEN FABRICS

Flannel uses either worsted or woollen yarns woven using a plain or twill pattern. A finishing process called 'milling' brushes the fabric to raise the fibre ends. This fabric is quite soft and is commonly used today for nightwear and babies' clothing.

Crepe has a crinkled appearance and is made using highly twisted wool or worsted yarns which allows it to spring and drape exceptionally well.

Tweed is well known and popular for its hard wearing woven woollen cloth. It is used for suits and coats.

KNITTING

Knitting is the second most important way of making fabric. It uses one piece of yarn and produces an elastic, stretchable fabric. Knitwear today is usually made on large knitting machines, but knitting is still a craft in its own right. We often see very creative original knitted jumpers.

Knitting is a quick and cheap form of fabric construction and can use either natural or synthetic fibres. In simple terms, knitting is the construction of fabric by forming loops of yarn with needles and drawing new loops through the previous ones.

Knitted fabrics have continued to increase in popularity with various fashion trends. Generally, knitted fabrics are very comfortable, crease resistant and allow body movement. They are elastic, breathe well and are light, yet warm. Knitted fabrics do not fray and don't require seam neatening.

There are two basic structures—weft knit and warp knit—which require different machinery and produce different types of fabrics.

WEFT KNIT

Weft knitting has a longer history and tradition than warp knitting and is generally more familiar. Weft knit requires a single yarn to travel horizontally making a row of loops into which the following rows of loops are knitted, like for hand knitting. Three basic stitches are used to make weft knitted fabrics: plain, purl and rib. Perhaps you are familiar with hand knitting and are already aware of these basic stitches.

Plain uses one set of needles, either flat or circular, and is used for jumpers, hosiery and underwear garments. It ladders easily unless specially constructed.

Rib uses two sets of needles allowing greater variety in fabric design, particularly if needles are placed at right angles to each another. Rib has greater elasticity width ways and is therefore used for socks and underwear as well as wrist and neck bands on jumpers

Purl is a combination of both plain and rib stitches, using either flat or circular bed knitting. Purl boasts elastic properties lengthwise and crosswise. It is used in children's wear, such as jumpers.

Weft knitting is capable of knitting flat, circular or tubular pieces or complete garments that only require one or two seams to complete.

Can you find illustrations using these different knitting styles?

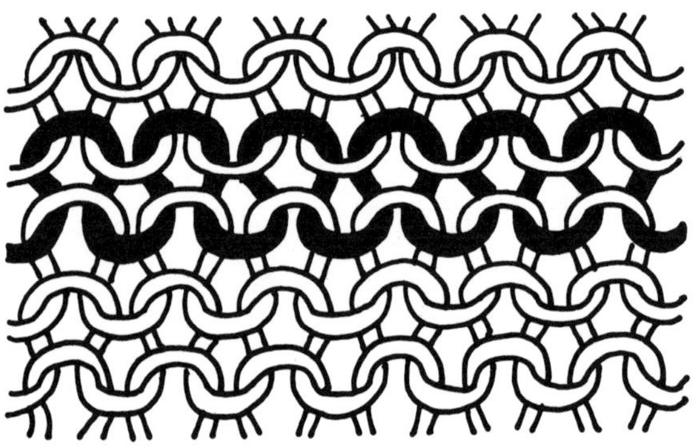

Fig. 1.11 Weft Knitting

WARP KNITTING

Warp knitting has only come into being since the 1960s and therefore isn't as well known as weft knitting. Warp knitting differs in appearance and method of manufacture. During warp knitting, several hundred to over one thousand yarns are interlooped vertically (warp direction) into adjacent yarns to form a flat fabric with straight edges. Warp knitting gives very stable knitted structures which are elastic and resistant to running and creases.

Examples of warp knitted fabrics include Tricot, which is used for underwear and other support garments, lingerie and swimwear. Cire (Nylon Tricot) is very shiny and smooth with a wet look surface often used to imitate leather. Warp knitting is often used for the construction of net or lace curtains, machine crochet and all kinds of lace.

Fig. 1.12 Warp Knitting

STUDENT ACTIVITIES

1 Choose a room in your house and list all the different types of the textiles you can see. Select five and explain their function—are they floor coverings, or clothes, for example.
2 As a class, think of all the different fibres you can. Look at the labels on your clothes for some ideas. Each member of the class should then choose one that they know little about or are interested in and answer the following questions.
Is it natural or synthetic?
When was it first used or invented?
Were there any events at that time, such as war or revolution, that may have influenced its development?
Who developed it?
What was it first used for?
What is it used for now?
What effect has it had on people's lives?
How is it made?
Summarise the information and give a five minute talk to the class. Use illustrations to support your presentation.
3 What is the name of the fabric made from flax?
4 What does carding do?
5 Which two natural fibres crease easily?
6 Which state produces most of Australia's cotton?
7 What are the two types of synthetic fibres?
8 What was the first synthetic fibre to be developed? Who developed it?
9 In weaving, in which direction do the warp threads run?

BACK TO THE FUTURE

The development of textiles has been influenced by a variety of factors, but in particular, technology. In this chapter we will look at some of the influences of technology on the textiles industry, some historical aspects and consider what the future may hold. We will only be able to look at these briefly, but you could find out more about them.

The focus of this chapter will be:

- **the market**
- **social factors**
- **the past**
- **the future.**

THE MARKET

The market of any industry has a direct influence on the type and amount of the products produced. In the textiles industry, the type of market has changed over the years. Initially, fabrics and household products such as furnishings would have been a major part of the market. Nowadays, more industrial areas, such as tyres and tarpaulins, and products for sport and leisure, such as sailing and camping, have a far greater part.

SOCIAL FACTORS

Society has changed in many ways. One of the more obvious changes has been the increased number of women in the workforce. This has led to the development of fabrics which are far quicker and easier to care for as families have less time to spend on these tasks. It has also meant that a whole new range of clothing for working women has been developed. These are just some of the consequences of one change in society. There are thousands more. You could investigate some of these. Ask members of your family about the changes over the last 10 or 20 years—you may be surprised at what they tell you!

THE PAST

During the eighteenth century there was increased importance placed on the home. This placed a focus on the textiles industry—and an increase in demand for home furnishings and different types of fabrics. Developments in equipment and therefore in fabrics followed. We will look at some.

In 1733 an Englishman by the name of John Kay introduced a new invention which was to lead to huge advances in the weaving industry. It was called the Flying Shuttle.

As improvements in one area occurred, there were, of course, improvements in others, as each directly affects the others. The Flying Shuttle meant that weaving was much faster, easier and more versatile as only one hand was required to throw the shuttle back and forth. This put pressure on the spinning industry to produce yarn more quickly to match the faster rate of weaving.

In 1763, James Hargreaves invented the Spinning Jenny. This was followed by Richard Arkwright's Water-frame Spinning Machine in 1769, and Samuel Crompton's Spinning Mule in 1779, which combined characteristics of the two. All of these machines led to vast improvements in the industry.

Fig. 2.1 James Hargreaves' Spinning Jenny

Dr Edmond Cartwright patented his first power driven loom in 1785 and a newer version in 1787. This gave the factories even greater output. One of the biggest advances in weaving took place in the French silk industry when J.M. Jacquard invented a loom that could automatically weave patterns. A whole new area of textiles was developed. The Jacquard Loom became public property in France in 1806. This was the first loom to mechanically produce tapestry-like textiles.

Developments were also made in carding. In 1748, David Bourn and Lewis Paul patented a carding machine that could separate cotton and short fibres of wool. By 1785, Arkwright had developed a machine which could card continuously.

Perhaps the most notable advances in spinning came around 1850 with the introduction of the ring spindle by G.E. Donisthorpe. This spindle rotated at high speed producing yarn much more rapidly and significantly improved the quality of worsted yarn.

All of these inventions led to a textiles industry which could produce far greater quantities of fabric in less time.

Today, the industry uses high-powered machinery to produce far greater quantities of finished products than ever before. Computers are a part of all stages of the process, from the initial development of the raw materials through to designing, production and marketing of the end product. Computers have affected the quality of the products now available. They have also meant vast changes in the availability of work in the industry.

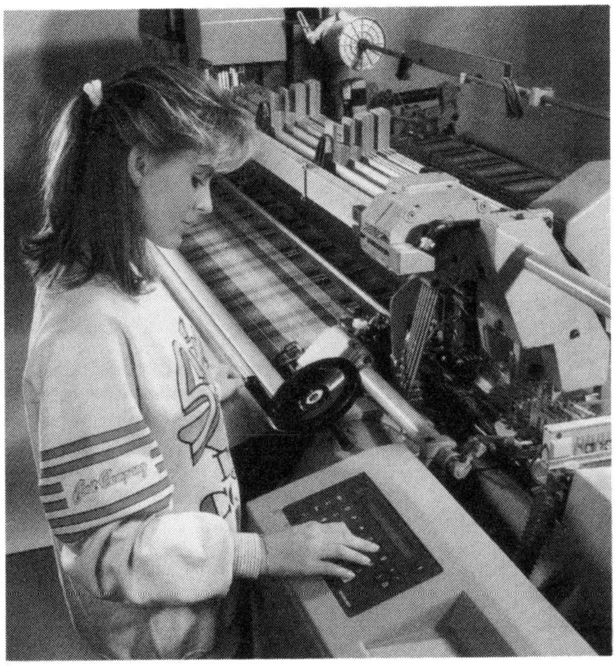

Fig. 2.2 A modern computer controlled loom

SEWING MACHINES

In 1851, an American by the name of Isaac Singer produced the first commercially available sewing machine. This was the first major domestic appliance to be sold to households. Prior to this machine, hand sewing was the only method used. Singer's sewing machine became very popular because it saved so much time, and the industry grew rapidly.

The first sewing machines were hand operated and only sewed forward. Now there are many different types of sewing machines which allow us to produce more complex and better quality products. We have computer-operated sewing machines that can do complex operations such as embroidery, monograms and circular buttonholes. This will lead to further changes in the area.

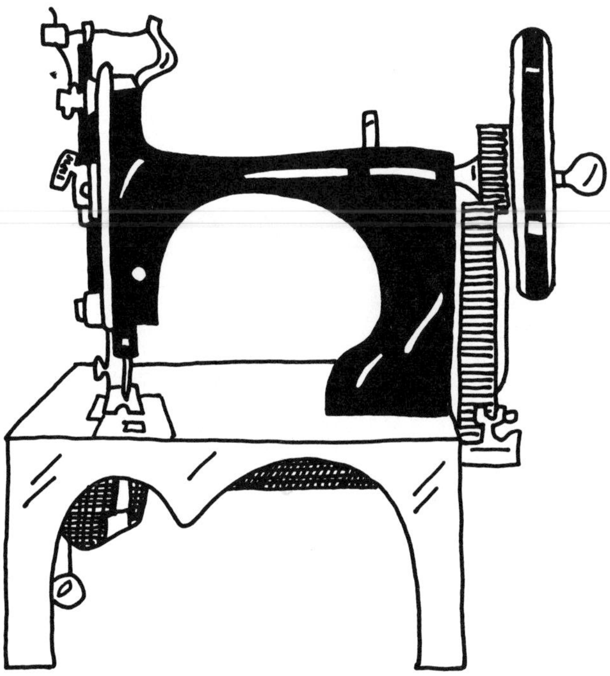

Fig. 2.3 Isaac Singer's first sewing machine, built in 1851

IN 2020

It is hard to predict what will happen in the future. People used to say 'I wonder what it will be like in the year 2000?' We are close enough now that perhaps the year 2000 is no longer such a mystery. But the year 2020 is. There will certainly be all sorts of changes by then due to such things as advances in technology, the depletion of natural resources and the ever increasing concern about the environment. We may not know exactly what these changes will be, but we can determine some of the things that will bring change to our lives.

These are only some of the issues. Let's have a brief look at them.

TRAVEL

By 2020, we may be regularly travelling in space. The textiles used for clothing, furnishing and in industry will have to change to suit our new needs. We can ask ourselves many questions:

What sort of clothing will we need in space?

What fabrics will be used in the seats for our travel vehicles?

HOME

What new fabrics have you seen in your home over the last ten years? There have been huge changes in the past ten years and we seem to be changing more rapidly all the time. Consider what a home in the year 2020 may look like.

Will we sit on chairs?

Will we have curtains or blinds?

What sort of floor coverings will we have?

Fig. 2.4 A modern computerised sewing machine

Fig. 2.5 Future influences

LEISURE AND SPORT

Our leisure activities certainly affect what clothing we wear and our needs in furnishing and industrial textiles. Consider what sports we may play in 2020. Also think about the huge changes in the type of clothing worn for sport—bathers were neck-to-knee garments in the 1940s! Synthetics have allowed us to have garments which are quick and easy to wash and dry, stretch, are bright in colour and light to wear. Do you think we will advance further in sportswear?

Our leisure activities have a large influence on our home life and general lifestyle. Huge advances in technology have reduced our domestic workload giving us more leisure time. Perhaps we will have even more in the future.

WORK

The type of work we do determines the clothing we wear. It can also affect many other aspects of our lifestyles as it determines how much money we earn. Consider the clothing worn by police officers and nurses. These are jobs in which the workers have uniforms that must be worn for safety, health risks and identity. There are also jobs which, although they don't have a set uniform, still call for a certain type of dress. (For example, a motor mechanic wears tough overalls while a shop assistant may need to look neat and tidy.) Consider some of the jobs which people may have in 2020 and the clothing which would be required for these jobs.

ENVIRONMENTAL ISSUES

Over the last few years in particular, we have become much more environmentally aware (mainly because the environment has become a political issue). This has influenced many aspects of our everyday lives and will continue to do so in the future as we attempt to make better use of our resources. It means that we will need to look at recycling our goods and cutting down on waste. This will affect us in hundreds of ways. There are many environmental groups fighting to create a greater awareness of our environment.

This chapter has merely touched on the many aspects of the development of textiles. It is a huge area. By working on the student activities you may be able to discover more about areas that interest you.

STUDENT ACTIVITIES

1 Select a fibre or fabric which has been developed over the last fifty years and look at its development. How and why was it developed?
2 Select one of the inventions mentioned in this chapter and find out more about it. Why did it come about? What influences it had on the industry?
3 What was the name of John Kay's invention? When did he invent it?
4 Who developed the first sewing machine? When?
5 List as many jobs as you can that are associated with the fashion industry. Select one that interests you and research it further.
6 Select a sport that you think may still be played in the year 2020. Design an outfit for someone playing this sport.
7 List the ideal qualities of a futuristic fabric that could be used in fashion, furnishing or leisure equipment.
8 Clothing is one use of textiles. List another three.

THE WOOL STORY

> Chapter 1 briefly discussed the many different aspects of textiles including a number of fibres and their properties. This chapter outlines the development of the wool industry in Australia since its humble beginnings in 1788 to the multi-million dollar industry it is today. It is important to recognise the great effect the wool industry has had on Australia's development over the past 200 years.
>
> In this chapter we will analyse the production of wool: from the paddock to your woollen socks. The focus of this chapter will be:
>
> - Australia's wool history
> - the farm cycle
> - trademarks of wool
> - manufacturing processes.

WOOL PRODUCTION

What do you know about wool? For more than 200 years, wool has been a vital part of Australia's economic growth and prosperity. Australia is the world's largest wool producing country and supplies 70% of the world's traded apparel wool. Almost 97% of Australia's wool production is exported, comprising 10% of the nation's export total.

Australia's wool history is a very rich one. One hundred years ago, it was said that Australia 'rode on the sheep's back'. Today, with more than 80,000 wool growers producing some 29% of the world's total production, wool remains a great Australian success story. Did you know that the first sheep arrived in 1788 with the First Fleet! However, these first sheep weren't very suitable for wool but were mainly used for food.

The real Australian wool story begins with the arrival of the first Merinos from Spain in 1797. These sheep were purchased for a mere £8 and later sold to various leading land holders—including Captain John Macarthur, who has been noted for his insight and vision. Do you have an old $2 note? You will find that John Macarthur has been acknowledged for his great contributions by appearing on our currency.

Like most industries, woolgrowers have experienced booms and depressions. The industry continued to flourish to become the truly international industry it is today. To find out more about the growth and

Fig. 3.1 Australia's $2 note was a tribute to John Macarthur

development of wool in Australia, find out what effect the following incidents had on the wool industry:

- Depression of 1840s
- Introduction of fencing
- Discovery of gold
- Wool exporting
- Development of transport systems
- The Labour movement
- The Great Depression of the 1930s
- The First and Second World Wars.

WOOL GROWING

Sheep can live in most of Australia apart from the central desert and tropical areas. Most are raised inland of the Great Divide. Looking at Figure 3.2, you can see that around 40% of the sheep population is in NSW. Western Australia, the second largest wool-growing state, produces around 20%. Australia has approximately 60,000 wool growing properties. On average, these farms carry 1700 sheep. Some carry much more: 1% of farms carry more than 20,000 sheep.

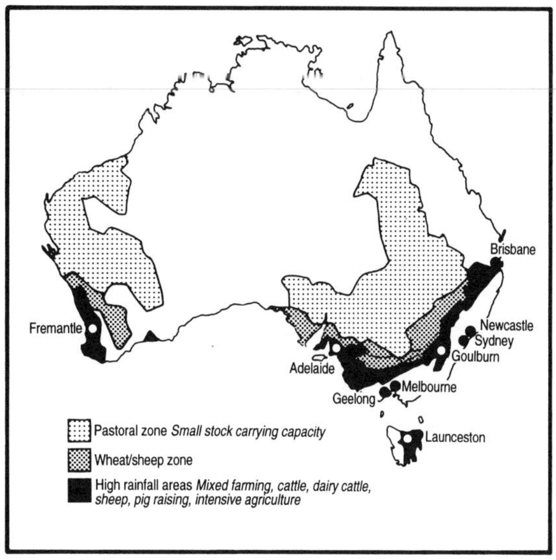

Fig. 3.2 Australia's wool growing areas

The activities on sheep farms vary from one part of Australia to another. These differences are due to different weather patterns and the preferences of individual farmers.

Use a calendar to summarise the 'typical' cycle of work which is generally common in most parts of NSW, Victoria and South Australia. Remember the cycle of activities can vary from one part of Australia to another and according to individual farmers and managers.

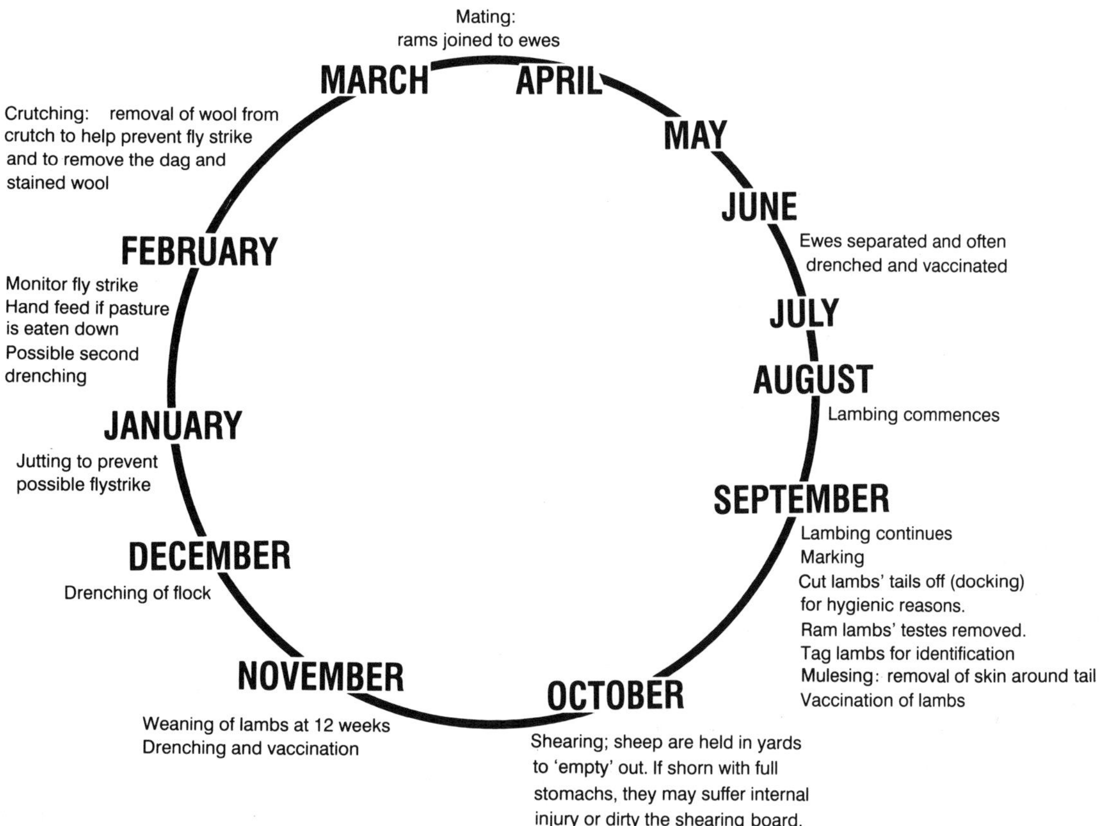

Fig. 3.3 The Farm Cycle

TRADEMARKS OF WOOL

During the 1930s, intensive research was undertaken into the production of synthetic fibres. The wool industry was feeling quite threatened and in 1937 formed the International Wool Secretariat (IWS) to increase the demand for wool. The IWS works on behalf of woolgrowers in Australia, New Zealand, South Africa and Uruguay, which account for 80% of the international wool trade.

Fig. 3.4 The International Wool Secretariat logo

The IWS doesn't make or sell wool products, but works with manufacturers and retailers to create consumer demand for quality wool products by identifying new market opportunities and developing new products. These activities include advertising, product development and design services, specialist advice on technology and fashion, market information and Woolmark standards control.

In the early 1960s, wool was lacking status and as a result of the poor demand and a massive increase in world wool production, prices slumped. To help maintain quality, an Italian—Francesco Saroglia—designed the distinctive Woolmark symbol for the IWS's promotion of wool throughout the world. This symbol for pure new wool is the world's best known textile trademark. It is licensed for use in more than 70 countries and is recognised by more than 95% of Australians.

Pure New Wool simply means that is has been processed straight from the sheep's back. This symbol guarantees that the product has passed specific tests for qualities such as strength, colour fastness, durability, and shrink resistance. Woolmark products must contain a minimum of 95% wool fibre. The remaining 5% may consist of non-wool fibres for decoration or technical purposes.

Fig. 3.5 The Woolmark symbol

Fig. 3.6 The Woolblendmark symbol

Due to developments in textiles, the IWS introduced the Woolblendmark symbol in 1971 to identify products made from a combination of not less than 60% virgin wool with synthetic fibres. The addition of other fibres to wool has distinct benefits to the consumer. For example, wool has good crease retention properties, enhances durability, and is lightweight. It was hoped that the introduction of this symbol would increase the overall demand for wool. Interestingly the Woolblendmark does not apply to products which it is felt threaten the existing pure wool market such as carpets, blankets and hand knitting yarns.

AUSTRALIAN WOOL CORPORATION

The Australian Wool Corporation (AWC) was established in 1973. Essentially, the AWC replaced the pre-existing Wool Board and Wool Commission. The AWC has three main functions:

- operate the reserve price scheme, which guarantees wool producers minimum prices throughout each wool selling season
- promote and encourage woolgrowers to adopt high standards in wool production, preparation and handling
- promote wool in Australia, including promoting furnishing textiles and fashion apparel.

AWC activities also include wool industry research and innovation.

Fig. 3.7 The Australian Wool Corporation logo

CSIRO

Because of the importance of the wool industry to Australia, wool growing and wool processing have been subjects of intensive study by the Commonwealth Scientific and Industrial Research Organisation, the CSIRO. Their research into aspects of wool growing such as nutrition, genetics, livestock management, harvesting and disease diagnosis and control, is undertaken by the Divisions of Animal Production, Animal Health, and the Australian Health Laboratory.

The Division of Wool Technology looks at wool as a fibre. Formed in 1988, the division's facilities include a full-scale wool processing mill, extensive laboratories and engineering and electronic workshops.

The primary aim of the research is to increase the world demand for Australian wool. Their specific objectives are to:

- improve the harvesting, marketing and transport of wool
- improve the operations required to form a textile from the raw product
- improve the performance of wool textiles
- develop technologies that enable new wool products to be manufactured

- reduce the environmental effects of wool processing
- increase our knowledge of the basic properties of wool fibre.

As you can see, there is a lot of research into both the animal and the processing of the raw material. Research by CSIRO's Division of Wool Technology has been responsible for a number of achievements which have had dramatic effects on the marketing, processing, and performance of wool and wool textiles. We will briefly mention a couple of these to provide you with a basis for your further research.

SI-RO-MARK SCOURABLE BRANDS

Tar brands once used on sheep could not be removed during scouring and had to be hand-picked from the fleeces before further processing. Si-Ro-Mark branding fluids are made from pigments which are removed during normal scouring, saving at least $3.5 million since the development of this fluid in the late 1950s.

SIROCLEAR

Siroclear is a device that detects the presence of contaminants such as burrs and seeds, dark fibres, baling twine and pack materials in undyed wool yarns. Once detected the contaminant is removed during the winding process. Previously, contaminants were eliminated by hand at the fabric stage, a costly process. Siroclear offers substantial cost savings to the producer of white or pastel wool fabrics, where dark fibres can be catastrophic.

SHRINK RESISTANCE

Several shrink resist techniques have been developed depending on the final use of the wool and the stage of processing required.

The Chlorine-Hercosett process is now widely used throughout the world, and is the basis of the International Wool Secretariat's Superwash labelling program for machine-washable wool.

Sirolan BAP is another simple process for making wool fabric shrink-resistant. The fabric is impregnated uniformly with a mixture of two synthetic resins and dried. In drying, the resins form bonds between fibres which prevents them from moving and hence the fabric from shrinking when it is washed. This process is now used throughout the world for apparel and other textile products such as blankets.

PHOTOPROTECTIVE TREATMENT FOR WOOL

Most textile fibres, including wool, change colour and lose strength when exposed to sunlight. Curtains and automotive upholstery are particularly prone to degradation. Photoprotective agents improve wool's performance in these areas. This research has resulted in improved durability and light fastness which will see wool penetrate further into the automobile upholstery, curtain and apparel markets.

MANUFACTURING

An incredible 97% of the wool produced in Australia is exported to wool textile manufacturers throughout the world. Most of the wool is exported in its 'greasy' form, that is, as it leaves the sheep's back. Wool must pass through many stages to be manufactured into textile products. In the following section we will travel with the wool through its various manufacturing processes.

Like anything, wool differs in quality for many reasons; the health of the sheep, where the wool was

Fig. 3.8 Shearing

grown, which part of the sheep the wool came from, and so on. Wool is therefore classed according to quality—this is the first process that raw wool goes through to ensure that the manufacturers obtain maximum spinning results from a fleece.

SCOURING

The fleece first passes through a series of steel teeth which separate the raw wool fibres. The wool is then washed repetitively in warm soapy water to remove dirt and grease. This process is called scouring.

CARBONISING

Straw, burrs and any other vegetable matter are removed by treating the wool with acid, heat and the pressure of rollers. This process, called carbonising, ensures that any remaining impurities are crushed and blown out of the fibre.

BLENDING

The wool is then blended to ensure uniformity of colour and quality. Many different wools may be blended together. For example—short fibres combed out during the worsting (described below) may be blended with scoured wool from the woollen system. To prevent the build up of static electricity, lubricants are added.

From this stage there are two main processes to convert raw wool into yarn:

- woollen system
- worsted system.

Woollen yarns are usually woven in plain or novelty weaves and worsted yarns in twill.

WOOLLEN YARN

The woollen system uses short fibres of different lengths to produce soft, lightly twisted, bulky yarns with protruding fibres that make them springy to handle and give a fuzzy or rough appearance. These yarns are not as durable as worsted yarns, won't hold creases, and are easily soiled and easy to clean. Woollen yarns are less expensive than worsted yarns.

Woollen yarn is used in woollen fabrics, such as tweeds, flannel and blankets, and knitting yarns.

WORSTED YARN

Worsted yarns are made from long staple fibres of the same length. They are fine, firm and highly twisted and give a smooth surface to fabrics. The fabrics are durable, hold creases well, soil very easily and are usually woven in a twill weave.

Examples of worsted fabrics are dress materials (firmer) and knitting yarns for socks. They are mainly used for suiting fabrics and other tailored materials that require a smooth finish.

Worsted yarns require many more steps in their manufacture. Figure 3.11 will assist you in spotting the different processes.

Figure 3.11 shows the sequence of processes in the woollen system—sorting, scouring, carbonising, blending, carding and spinning—and the worsted system, which includes extra processes—gilling, combing, and drawing. Because worsted yarns require more processing, they are generally more expensive than woollen yarns.

CARDING

Carding is where the wool is fed between wire 'brushes' to form a continuous thin untwisted wad of wool. In the woollen system, the final section of the carding machine divides the wool into strips to form slubs, which are wound onto bobbins ready for spinning.

When making worsted cloth, there are additional procedures: gilling, combing and drawing.

GILLING

Gilling aligns the different blends together by passing them through a series of coarse-toothed combs. This prepares the wool for combing.

COMBING

Combing is a continuation of the gilling process but uses a fine-toothed roller comb. This process is crucial as it removes any remaining impurities and any short, broken fibres from the sliver. These short fibres, called noils, are then used in the woollen system. The remaining wool is then wound into balls called 'top'.

DRAWING

Drawing divides the 'top' further into a suitable thickness ready for spinning. This is usually repeated several times to reduce the top and add a small amount of twist to strengthen the fibre. The reduced sliver is now called a roving.

SPINNING

Woollen and worsted fibres are both drawn out between rollers and twisted into a thread. Woollen yarns are not twisted as tightly as worsted yarns. The yarns are then wound onto large reels ready for any of the final processes. Spun yarns may be used as a single strand or twisted together (plied) with others depending on how they are to be used. Think about hand knitting yarns, which are often a number of strands combined.

Technology has contributed to some major developments in spinning, including the Sirospun yarn spinning system which operates all over the world. Since its release in 1980, more than 150,000 spindles worldwide have been fitted with Sirospun. The process is especially suited to the production of lightweight, trans-seasonal, 'Cool Wool' type fabrics. The process can be performed with only minor changes to conventional ring-spinning equipment, and substantial cost-savings result. Perhaps you can find out more about this process.

DYEING

Dyeing colours the wool and can be performed on scoured wool, top, yarn, fabric or the garment itself, although it is usually the yarn which is dyed. Piece dyeing of cloth is generally a cheap method of colouring fabric. After weaving, the material is dyed in

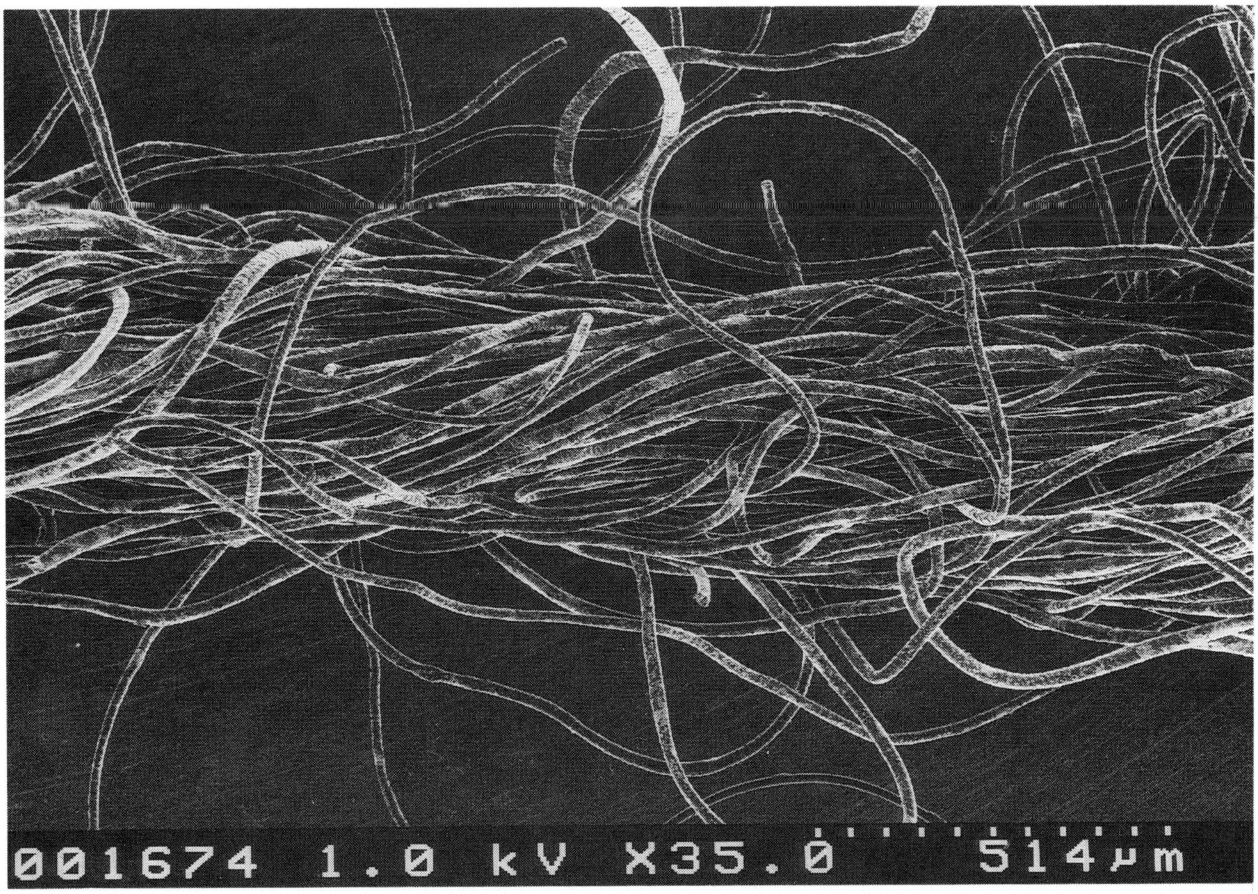

Fig. 3.9 Woollen yarn

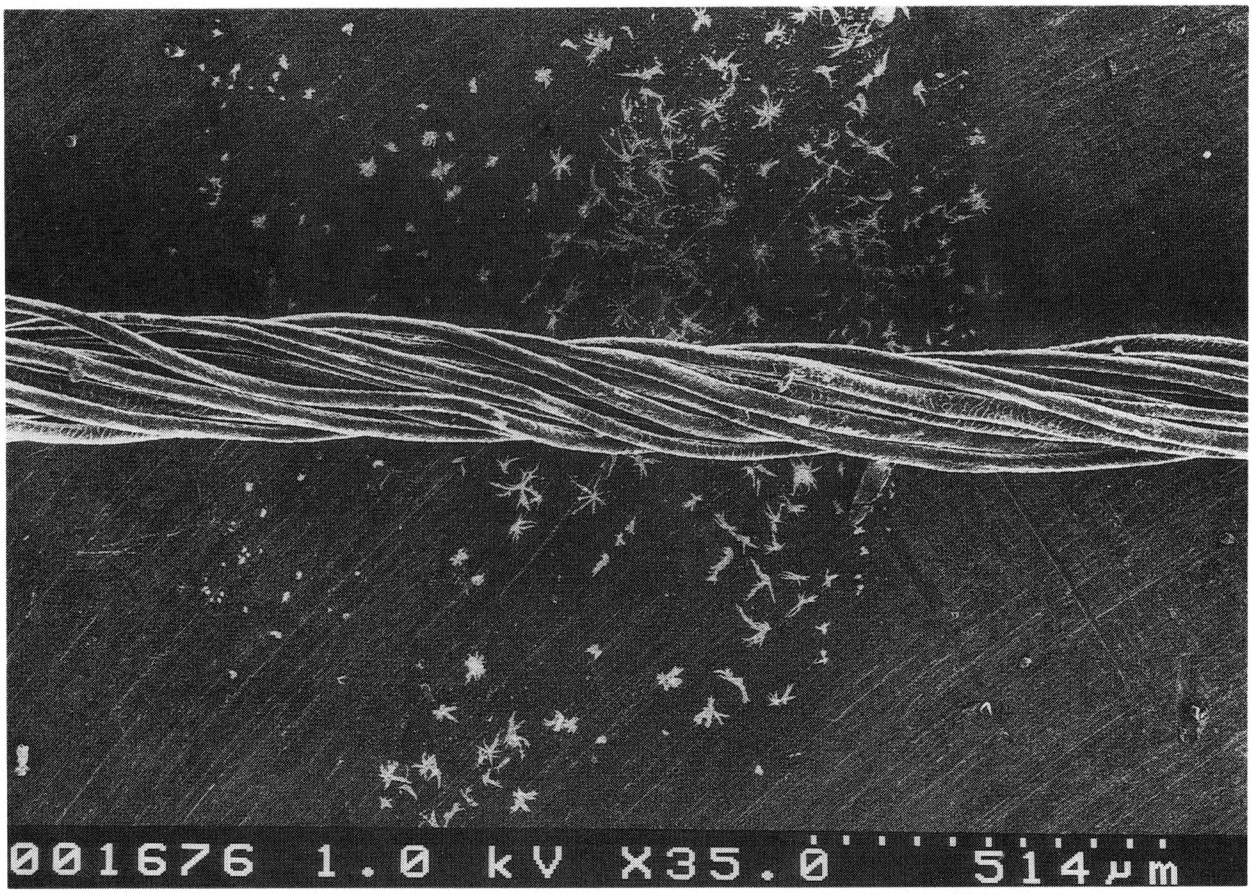

Fig. 3.10 Worsted yarn

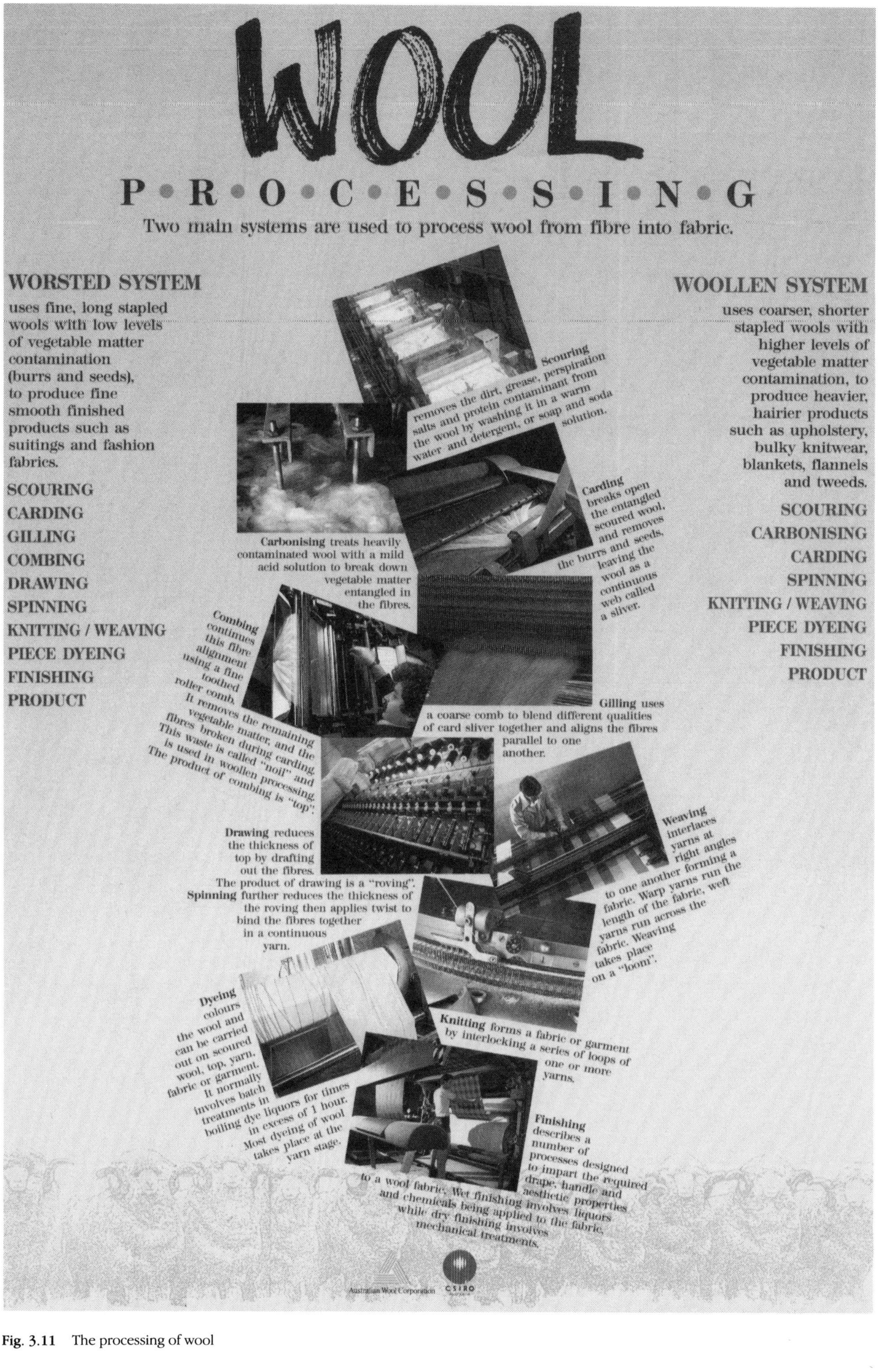

Fig. 3.11 The processing of wool

a piece which ensures even surface colour. Frequently though you will discover that the dye hasn't completely penetrated the threads. If you cut the edge of the fabric, it may reveal undyed fibres underneath.

PRINTING

Another way to colour wool is through printing. With printing, the colour is applied in a pattern. Normally wool needs to be treated with chlorine before printing. The dyes are applied through flat or rotary screens, or transferred to a fabric from special papers by hot pressing.

FINISHING

Finishing collectively describes a number of treatments designed to stabilise the shape, feel and other properties of the wool. Included in finishing are processes such as milling, scouring, pressing, raising and cropping.

AUSTRALIAN TANNING INDUSTRY

Australia produces over $600 million worth of cattle hides and sheepskins each year and is one of the world's largest exporters of these commodities. Of the 7 million cattle hides produced, some 2 million are tanned in Australia with the balance being exported after preservation by salting. About 90% of our 35 million sheepskins are exported unprocessed.

The Australian sheepskin tanning industry processes about 3 million skins a year for footwear, rugs, car seat covers and bedding underlays. Hides and skins are valuable primary products. On average a sheepskin is worth 30% of the value of the sheep. A major aim of the hides, skins and leather research program of the Division of Wool Technology is to enhance the technology of hide and skin processing for the benefit of the Australian industry. This is an interesting aspect of textiles research. You may wish to trace the major achievements in this area and investigate directions for a stronger future in hide processing technologies.

STUDENT ACTIVITIES

1 Look at the production of wool, including farming and fabric processing. Show all the major developments of wool and sheep in Australia since 1788 on a calendar of events. What influence has technology had on its growth and development?
2 Obtain some 'raw' wool from a craft shop and try spinning, weaving or knitting.
3 Look at some textile art or furnishings and identify those using some hand-made techniques. For example, lamp shades, scarves, rugs, and knitted jumpers. What are the advantages and disadvantages of each? Discuss the possibility of technology taking over all 'craft'-like industries.
4 Experiment with different dyeing techniques to determine their suitability for different applications.
5 Collect samples of various fabrics illustrating different weaving and knitting styles. Carefully label each one and record the following. (This record will also help you when you start material testing activities.)
   - Name of fabric
   - Fibre content of fabric
   - Characteristics such as weight, crease resistance, warmth, tactile characteristics (feel)
   - Possible uses and suitability for a given situation.
6 Find the meaning of:
Blending
Noils
Roving
Milling
Drenching
Apparel
Lubricants
Mulesing

MAKING PROJECTS FROM FABRIC

This chapter will give you an opportunity to put all of your knowledge and skills into action and create your own designer jumper. First we identify some simple and effective decoration techniques you may wish to experiment with. Some handy tips follow on garment construction which should reinforce your previous knowledge. The exercises at the end of the chapter will give you and your friends an ideal opportunity to share ideas and start production.

This chapter will focus on:

- decoration techniques
- seam construction
- zips
- windcheater production
- special tips.

WINDCHEATERS

Loose-fitting windcheaters have certainly become a fashion statement in the 1990s. Some leading fashion designers have taken advantage of this and charge exorbitant prices for quite simple designs which display company logos or decorative emblems. We hope you find this following section helpful and are inspired to create your own original tops using some of these simple techniques.

The most basic windcheater starts as three basic pieces—front, back and sleeve. With some basic pattern making hints and using your creativity and basic design techniques outlined in Chapter 6, you will see how easy it is to create your own designer top.

The following section outlines some popular decoration techniques which can be used to brighten the plainest garment.

APPLIQUÉ

Appliqué, or applied work, is a wonderful technique for brightening up any garment. Quite simply, a motif or shape is cut from one piece of fabric, placed onto another and stitched into place. The stitching can be decorative, hand worked or simply stitched by machine. Traditionally, appliqué is sewn with a machine satin stitch but there is nothing to stop you using any of the many other stitches on your machine. Machine stitching is certainly harder wearing and

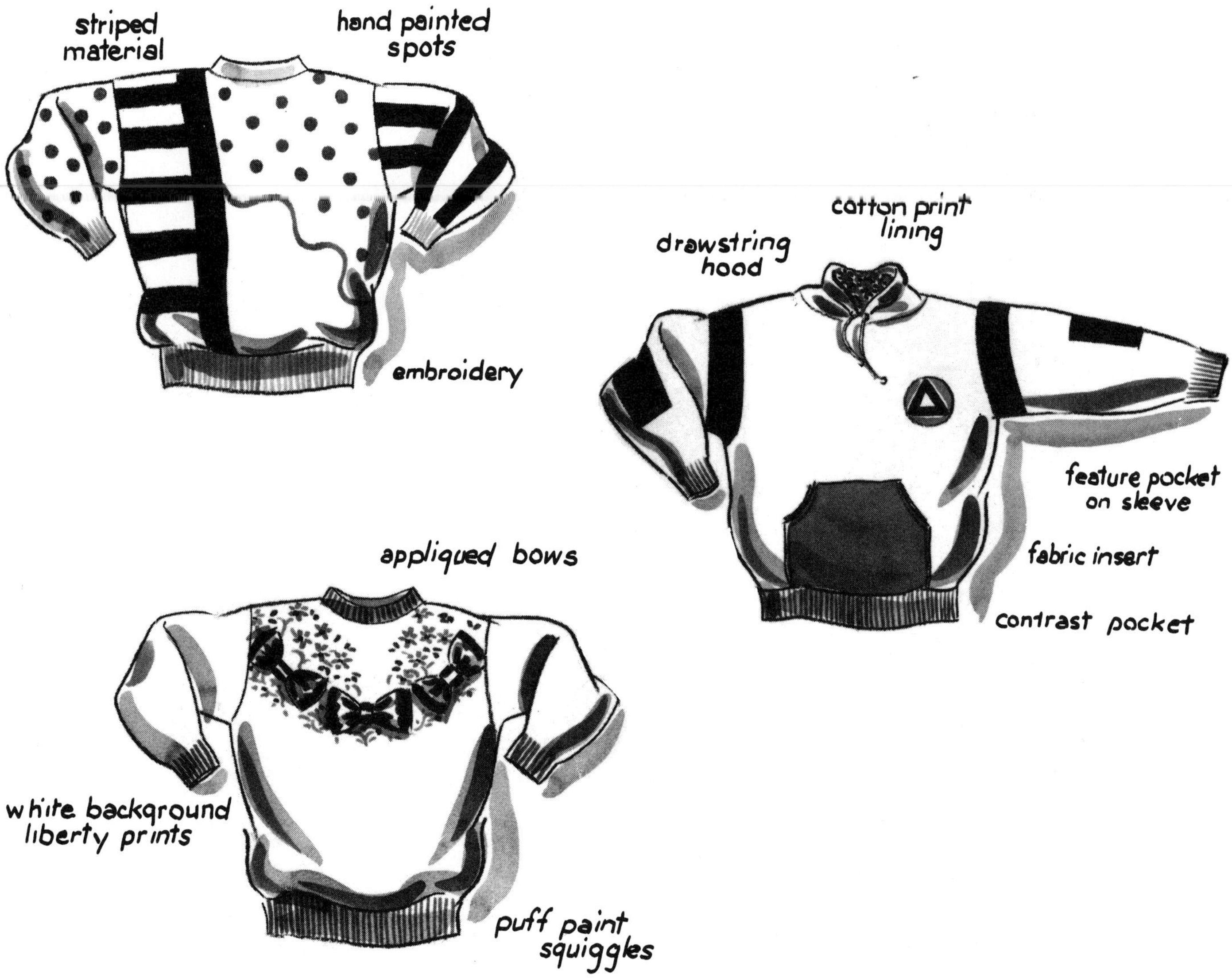

Fig. 4.1 Ideas for your windcheater design

Fig. 4.2 Appliqué design ideas

probably better for such wash and wear garments. Choose plain or patterned fabric and cut your motifs on the straight grain where possible to avoid stretching. Unless you are adding them to an existing item, apply the motifs before making up the garment. Firstly, cut a paper pattern for each shape. Pin the paper pattern onto the fabric and cut it out. No seam allowance is necessary. Vilene or 'magic paper' is excellent to use when sewing appliqué. You simply cut out the shape and then iron the shape onto the garment. When heated vilene acts as a glue and saves tacking the shape into place.

Next, using a clear plastic embroidery foot on your machine and matching or contrasting thread, stitch around the outside edge using satin stitch to cover the raw edges. Watch that most of the fabric is on the applied fabric.

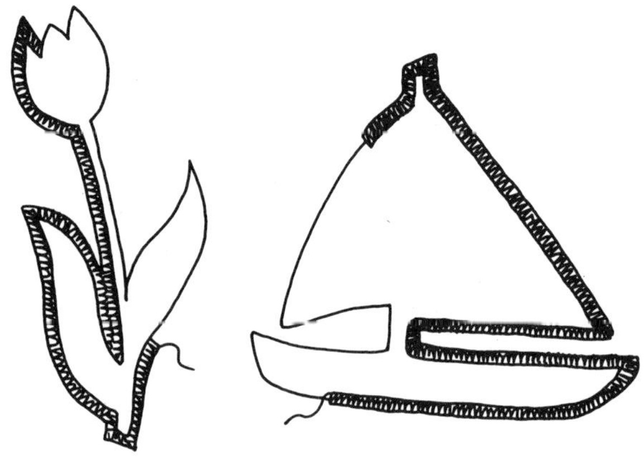

Fig. 4.3 Sewing appliqué

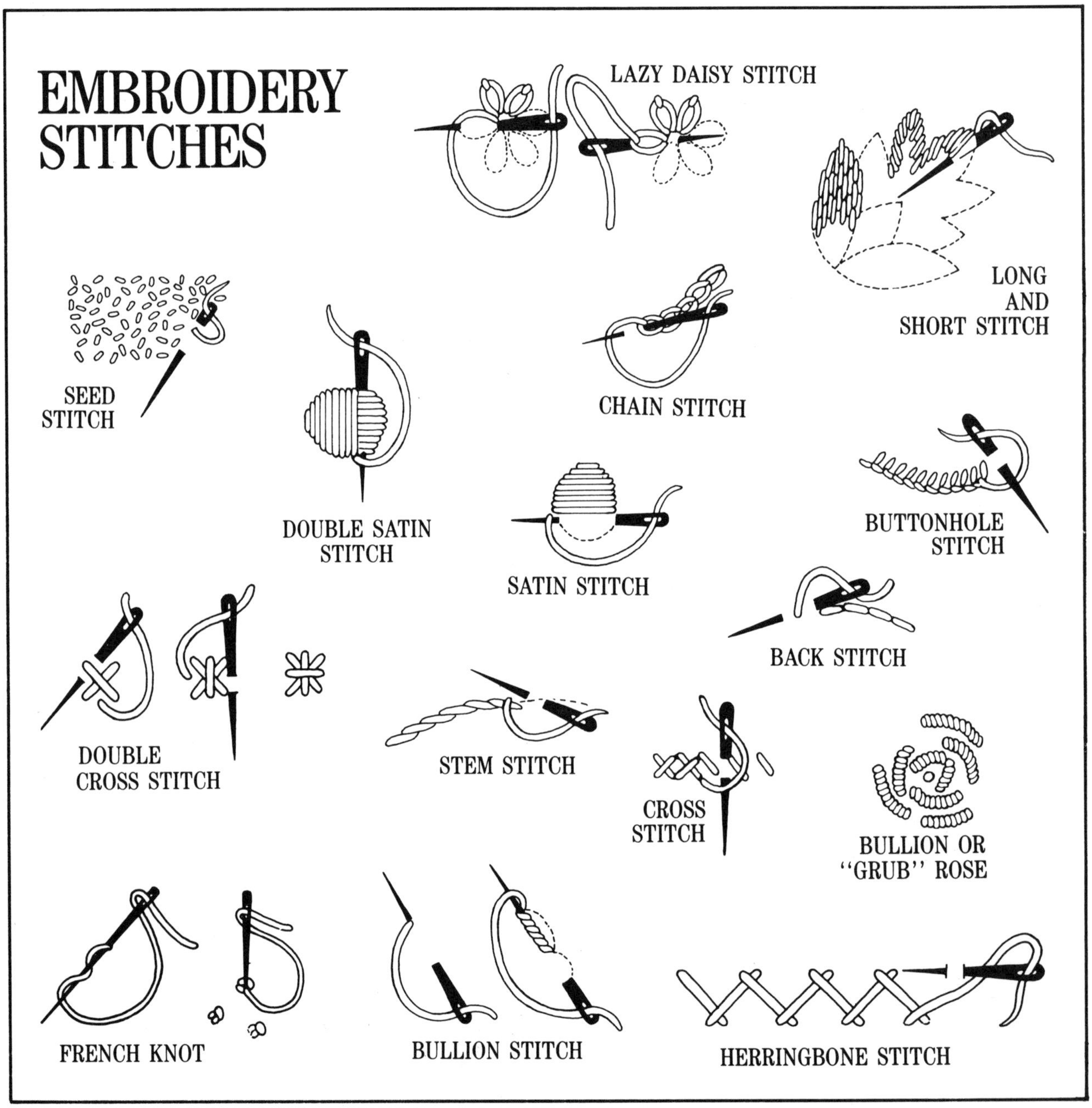

Fig. 4.4 Embroidery stitches

TIE DYEING

With the current 60s and 70s revival we have seen lots of colourful tie-dyed t-shirts and casual shirts. This form of dyeing provides a wide variety of colours and designs by simply tying parts of the fabric very tightly with string or rubber bands. You then dip the fabric into different dyes. The dye will only penetrate the fabric which isn't tied. You can either dye your fabric before you cut out your garment or after it is completed.

PRINTING

There are a number of exciting options currently available to choose from if you wish to print your own jumper.

Hand printing, using permanent fabric markers, squeeze bottles, brush and puffer paints in a wide range of colours, allows you to be very adventurous and create some wonderful designs. Permanent fabric markers allow you to 'write' or do fine line work. Think about using beads, sequins and other pieces to highlight areas.

Silk Screen is good for hard edge designs, particularly for silhouettes. Design your own personal label, school logo or a motif for a local sports club's uniforms.

Another decorative option is embroidery (either hand stitched or machine worked). You can create some lovely textures and maybe find the gentle arts quite relaxing while watching television. Figure 4.4 illustrates a number of popular easy to work stitches.

Batik is another fun way to create original fabric designs. Beading and sewing on buttons and odds and ends can create some wacky effects. Next time you're in the fabric shop or department store, have a good look at all the notions and sew-on trims available. These can supplement your other designs or be used individually. Try using trims, lace, fringing and badges. Cording and piping is a technique which is very effective in highlighting seams and creating exciting finishes.

SEAM CONSTRUCTION

Take a close look at the clothes you are wearing—the line of stitching joining two or more pieces of fabric is called a seam. It is important that you are aware of the various types of seams, their application to different fabrics, and their desired effect.

Before you choose which type of seam to use you need to consider the following points:
What type of garment is being made?
Is the fabric light and fine or heavy?
How will the fabric stand wear and tear?
Is the fabric likely to fray?
Is the fabric woven or knitted?
Do you wish to emphasise the seam lines?

The following section illustrates some basic seam techniques you will most probably use. If you require additional techniques, ask your teacher or refer to any sewing book.

FLAT SEAM

A flat seam is the most common and basic method used to join two pieces of fabric. Place right sides together and sew 1–2 cm in from the edge making sure that you stay exactly the same distance from the edge along the entire seam. Use a straight stitch on woven fabrics and a stitch that allows the fabric to stretch slightly for stretch fabrics, such as overlocking (four thread), narrow machine zig-zag stitch or machine stretch stitch. You will need to finish the raw edges to prevent fraying after you have sewn the seam.

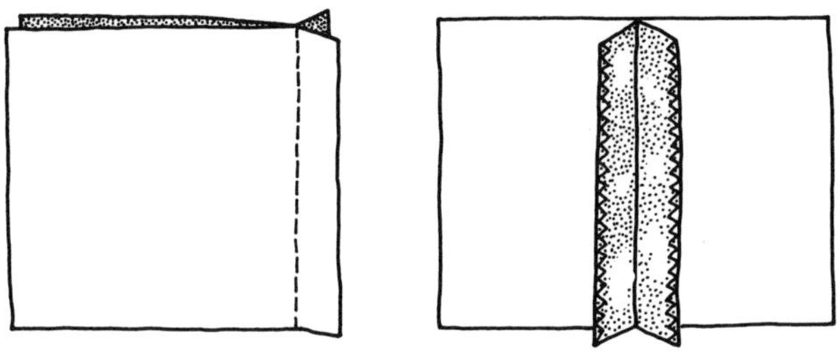

Fig. 4.5 Flat Seam

Place two pieces of fabric, right sides together, edges even, pin in place.

Machine stitch along seam line. Press seam flat.

FRENCH SEAM

A French seam is usually used on fine, see through fabric that is likely to fray badly. You usually use French seams for baby clothes, lingerie and some evening garments. French seams should be kept very narrow (no wider than 5 mm).

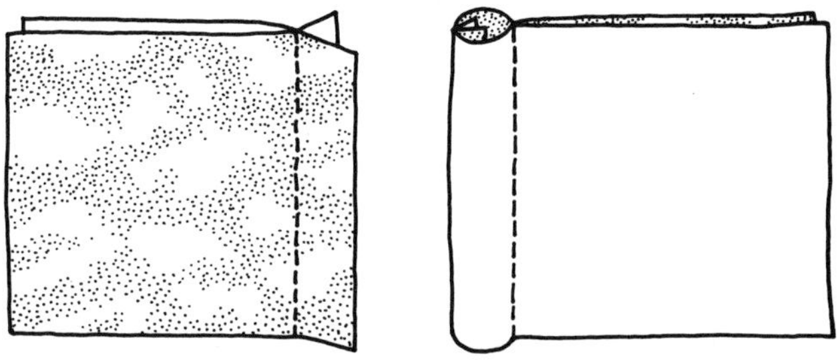

Fig. 4.6 French Seam

Place two pieces of fabric wrong sides together. Stitch about 1 cm from the raw edges and trim both seam allowances to 3–5 mm. Fold the fabric over so right sides are together and ensure the seam is 'rolled' to the edge. Pin and stitch 5 mm from the edge.

FLAT FELL SEAM

This is a useful neatening seam which is very secure and will not allow the fabric to fray. It is ideal for medium weight fabrics such as cottons and linen but can be too bulky for heavier fabrics like side seams for a denim skirt or tailored linen shorts. Sew this seam in contrasting thread for an interesting feature.

Place the two pieces of fabric wrong sides together. Pin and stitch about 1.5 cm from raw edges. Trim back one seam allowance to 5mm. Carefully fold the wider seam allowance in half, using the raw edge as a guide. Press the seam and carefully tack in place. Stitch along the fold.

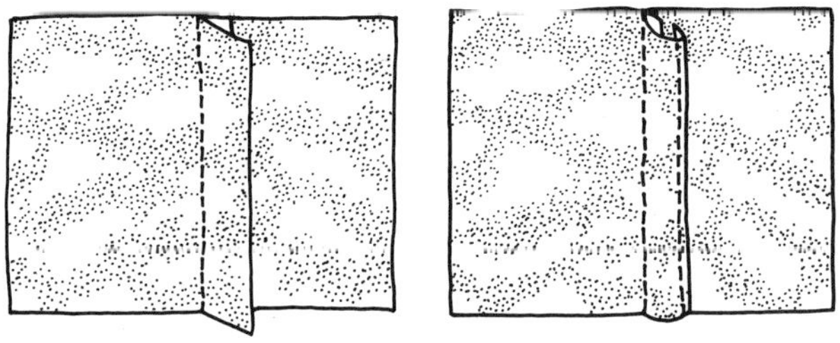

Fig. 4.7 Flat Fell Seam

TAPED SEAM

A taped seam is really an extension of the flat seam and should be used in stress areas or where the garment is likely to 'stretch' out of shape, for example, the shoulder seam of a t-shirt or windcheater.

Fig. 4.8 Taped Seam

Place two pieces of fabric, right sides together. Pin and tack some seam binding or tape along the seam line. Stitch together carefully with the tape side facing down.

An alternative method is to check whether seam tape can be threaded into the foot of your overlocker so it is sewn simultaneously with the seam.

PIPED SEAM

Piping is a very popular decoration technique in both fashion garments and home furnishings. Think about using a matching or contrasting colour or even a printed fabric. A piped seam can be either soft or corded—depending on the effect you want. First, cut 1.5 cm bias strips of the piping fabric and join them together to form a continuous strip or you can use commercially prepared bias binding.

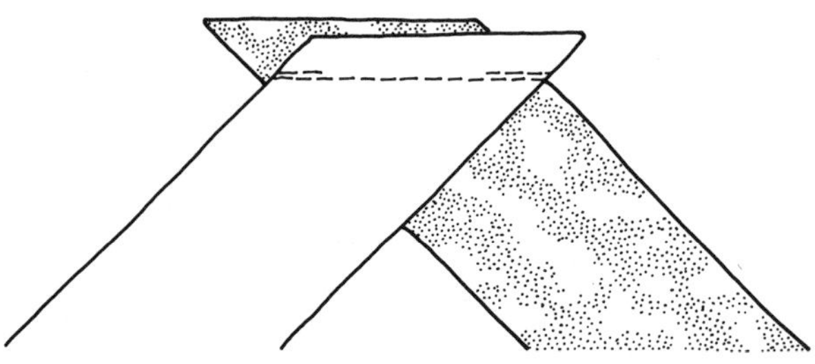

Fig. 4.9 Joining Bias Binding

Fold the strip with wrong sides together. Place the folded strip between the two layers of fabric (right sides together). Ensure all edges are even and stitch through all thicknesses. Trim away excess piping at the ends of the seam.

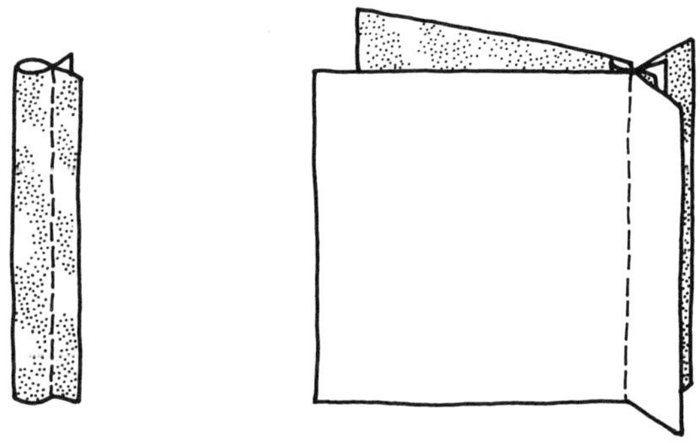

Fig. 4.10 Piped Seam

CORDED PIPING

Use the same technique as piped seam except fold the bias strip over cotton piping cord. Choose the thickness that you wish and adjust the width of the strip accordingly.

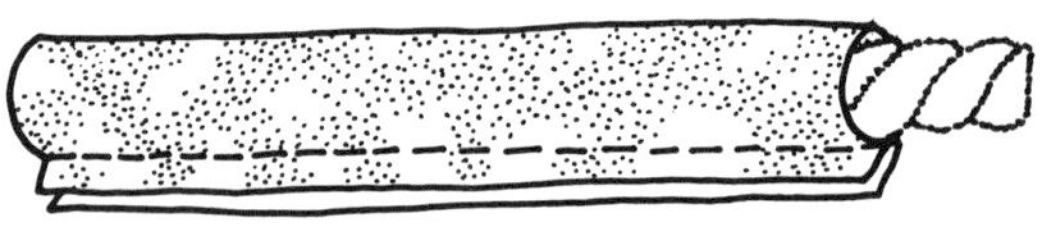

Fig. 4.11 Corded Piping

Stitch the piping first using a zipper foot as close to the cord as possible. Continue as with the piped seam.

ZIPS

Zips are more than a functional fastening used in pants and skirts; we now see the addition of zips to jumpers, tunics, and jackets for pure effect. Here are tips on basic zip insertion.

Zips are made in a wide variety of weights, materials and colours. As a rule, use a nylon zip on synthetic fabrics as a cotton tape may shrink during washing. Open-ended zips are often made of heavy nylon and will not rust from frequent laundering. There are a number of types of zips. The following are three standard techniques you are likely to use.

CENTRED ZIP

Centred zips are placed in the centre of a seam with equal amounts of fabric on both sides.

Sew the seam up to where the zip will begin and ensure you reverse five or six stitches to ensure strength at the stress point of the end of the zip. Continue to sew the remaining seam using stitch length 4 or 5. This will make it easier to unpick after the zip is inserted. Press the neatened seam flat. Pin the zip in position, ensuring that the teeth are centred over the seam.

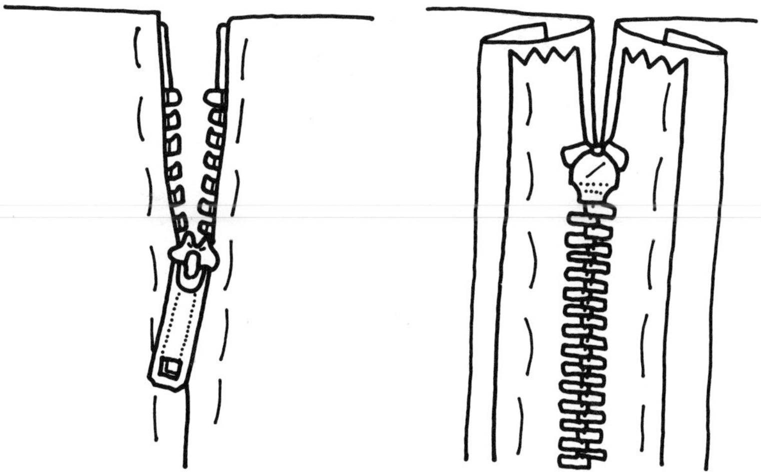

Fig. 4.12 A centred zip

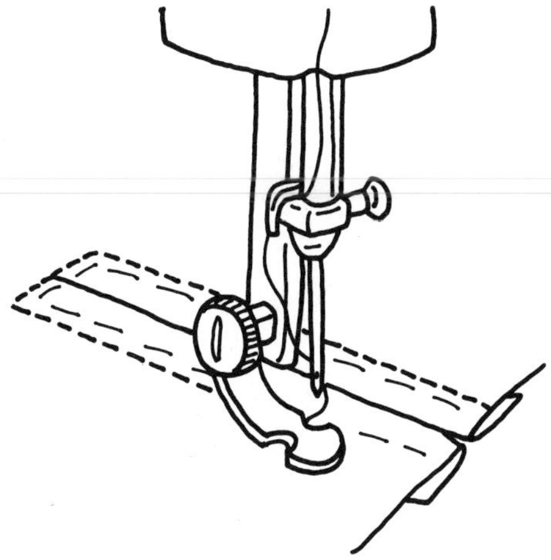

Fig. 4.13 Using a zipper foot

Tack 5mm from the zipper teeth and remove all the pins. Using a zipper foot, stitch close to tacking on the right side. Start at the top and stitch down one side, pivoting at the corners and stitching back to the top. Remove tacking.

OPEN-ENDED ZIP

Open-ended zips are usually centred with the teeth concealed like a standard centred zip. For a decorative effect, why not expose the teeth.

Firstly ensure that the raw edges are neatened. Pin and tack the seam allowance together along the fold line. Carefully pin the zip into position, either with the teeth centred over the seam or with the teeth exposed. Tack into place 5mm from zip teeth and remove pins.

Using a zipper foot, stitch close to tacking but sew both sides in the same direction. Remove tacking.

LAPPED ZIP

A lapped zip lies behind a flap formed by the seam allowance on one side. Usually you would use a lapped zip on the side seam of a garment. For example a skirt or trousers. The flap usually faces towards the back of the garment.

Firstly ensure that the raw edges are neatened. Press the seam flat and then, working from the right side, place the zip under the seam opening. Pin and tack one edge close to the teeth of the zip.

Lap the opposite seam allowance over the zip teeth. Make sure that the teeth cannot be seen. Pin and tack in position 1 cm from the fold. Stitch by machine as for a centred zip.

SPECIAL TIPS

Wherever a seam curves, it should be trimmed to allow the seam to fall neat and flat. In some cases, you will be required to grade the fabric, particularly heavier fabrics, to eliminate unwanted bulk.

NOTCHING

On the inside of a curve, always cut out small Vs, close to the stitching.

CLIPPING

On the outside of a curve, cut into the seam allowance, close to the stitching.

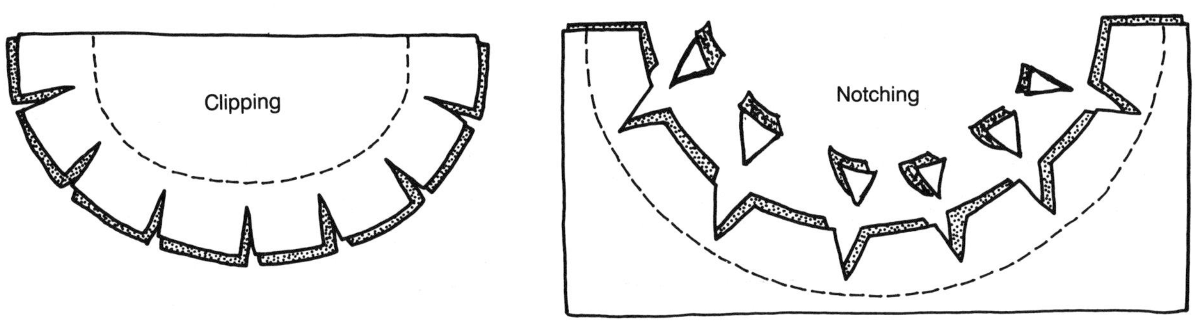

Fig. 4.14 Some sewing tips

GRADING

When bulk needs to be eliminated, such as on waistbands, cuffs, and collars, you will need to grade the seam allowances. Trim interfacing close to the stitching, then the under layer approximately 3mm, and the outer layer no greater then 5mm off the stitching line.

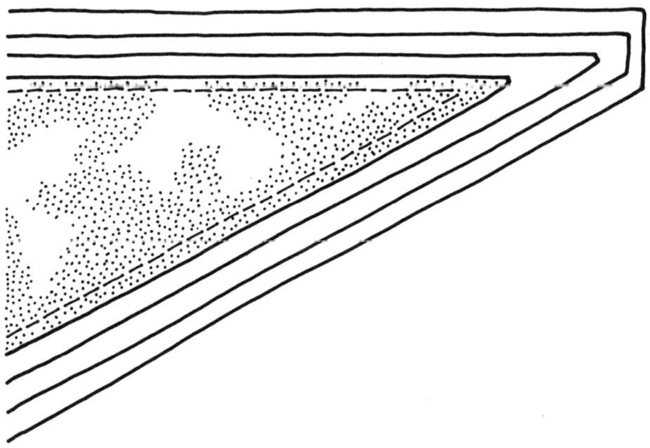

Fig. 4.15 Grading

MAKING A WINDCHEATER

We have suggested some possible ways you might wish to decorate a windcheater and hopefully you will now have lots of brilliant ideas of your own. If you're still a bit stuck, go window shopping or flip through some magazines or catalogues for some inspiration.

As a class, collect pictures and catalogues to get an idea of current trends and colours. For example, is piece work the go at the moment, or logos and cording? What about necklines—are they round, polo, V-neck? Do they have hoods attached? Are jumpers fitted or are they large enough to be called one size fits all? Are there different designs for boys and girls or are they androgynous? What is the basic theme this season? Can you identify lots of different trends happening?

Using these ideas, put together ten rough ideas of your own. Remember the design brief and all the different decorations you can use. Your teacher should be able to see diversity in your range and your ability to experiment with different techniques and design features. Your rough design sketches should be accompanied by lots of supporting notes.

Chat to your friends and share your ideas. Maybe they will have some suggestions for improvement. Discuss these designs with your teacher to clarify any loose ideas.

Choose the design you wish to make and render it. You will need to make a list of the processes involved and estimate the materials required. Start work on your design plan. You will need to carefully time each production process so it can be completed before the due date. It is a good idea to have some extra time up your sleeve at the end in case you have some unforeseen difficulty.

Prepare your pattern using the following tips to assist you.

Use McCall's pattern number 5606. We chose this pattern as it is a basic 3-piece pattern (front, back and sleeve), but can also be made with a hood and front pocket pouch. Trace the pattern onto a large sheet of

Fig. 4.16 Some design ideas for windcheaters

paper. Butcher's paper is fine. Clearly mark the front, back and sleeves. Now draw your design lines onto your pattern. Clearly mark each piece as you go.

You need to add seam allowances to all new design lines so trace around each piece again and add 1.5 cm to each new seam line. You don't need to add extra seam allowance to side seams as they are already included in the pattern. Make sure you clearly mark each piece, including grain lines, and add construction notches to assist you during the sewing of your garment. Carefully cut out each piece.

Now that you have your pattern pieces finalised you should be able to make a more exact estimate of the amount of fabric required. Purchase all your materials including matching thread and other notions.

Develop your work plan. Using the commercial pattern instructions as a guide you should not have much difficulty adding your own construction notes and processes. Make sure you indicate what sort of seams you will use. You may need to prepare some samples before starting.

Carefully construct your garment. Be sure to record your progress and any changes that you make to the initial design.

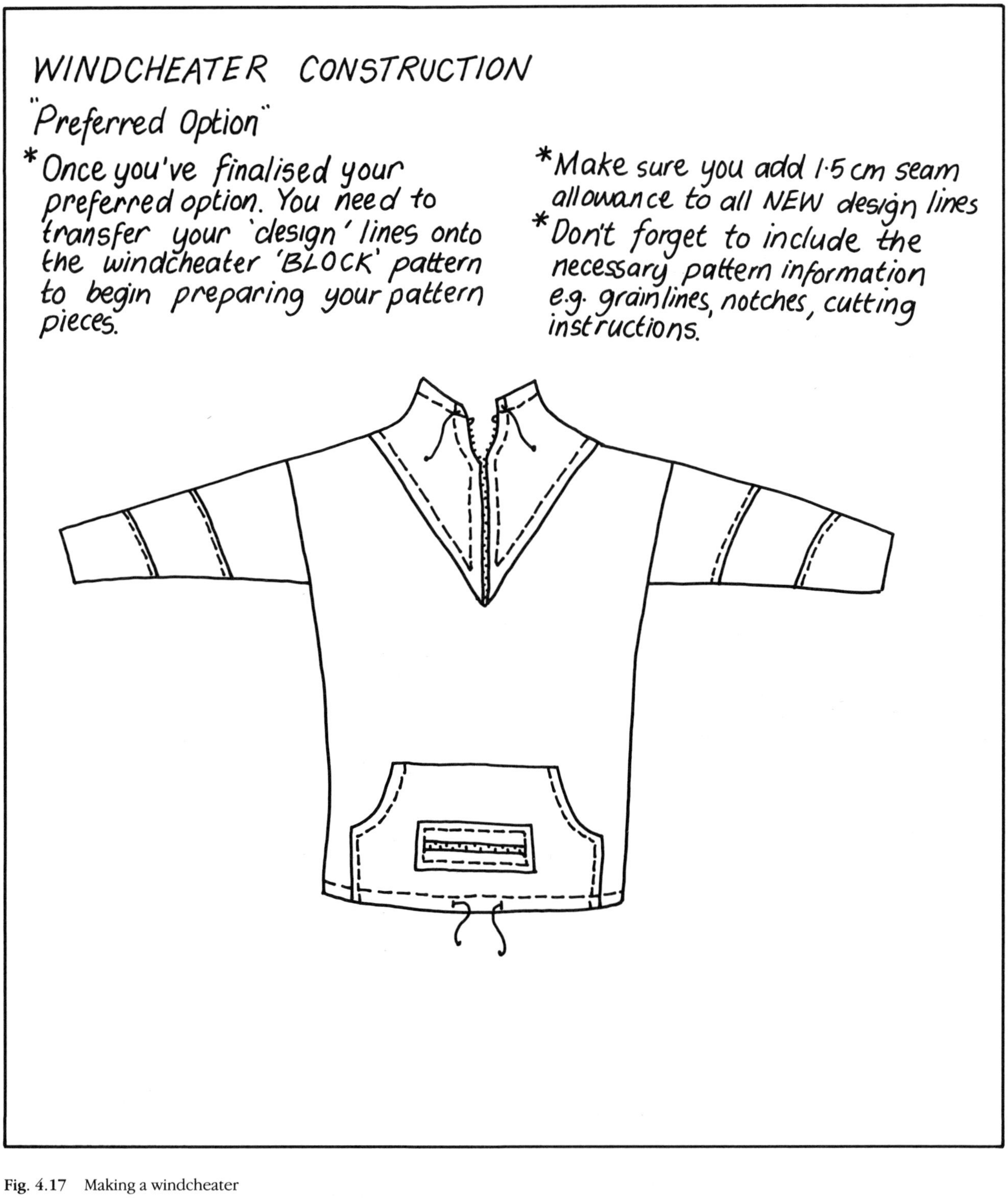

Fig. 4.17 Making a windcheater

WINDCHEATER CONSTRUCTION (cont'd)

"Working Drawing"
NOW make a list of tools, materials and processes required to complete your project.

TOOLS
- sewing machine
- overlocker
- pins
- stretch needles
- scissors

MATERIALS
- 1·3 m main fabric (150 cm wide knitted fleecy)
- 0·5m for feature fabric
- 2 zips (25 cm and 15 cm)
- matching and contrasting thread for top stitching
- cord for neck and waist

PROCESSES
- straight seam
- construction
- overlocking
- top stitching
- centred zip
- buttonholes

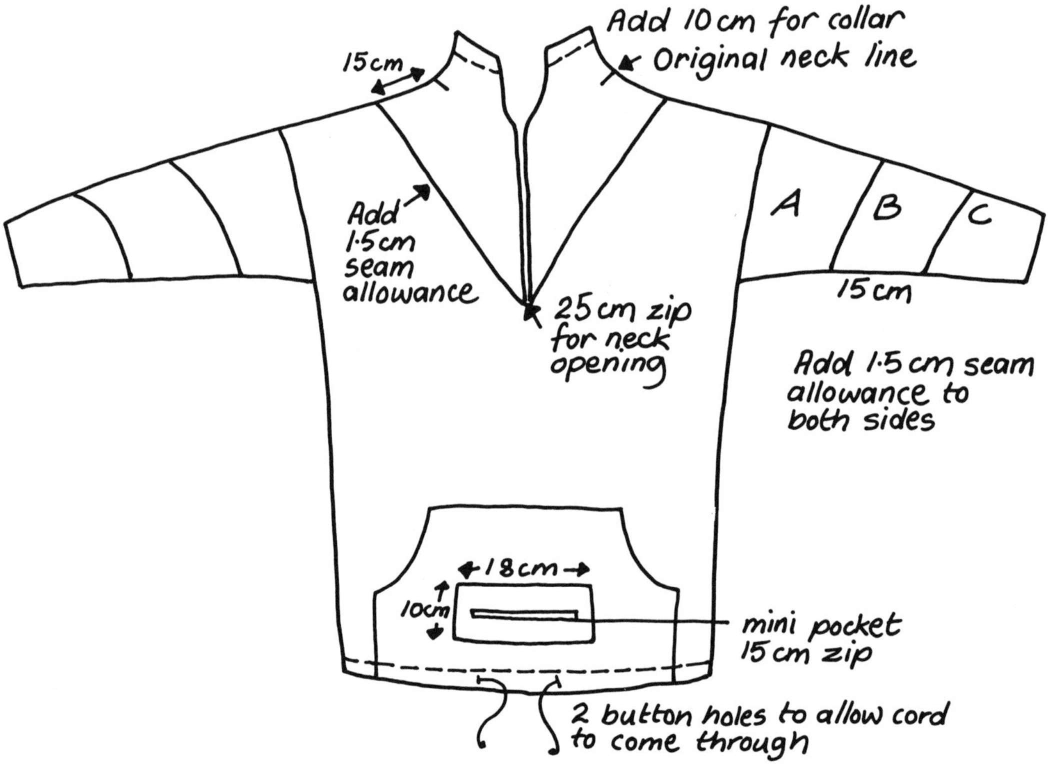

WINDCHEATER CONSTRUCTION (cont'd)

These are the pattern pieces required:

FRONT COLLAR - cut 2

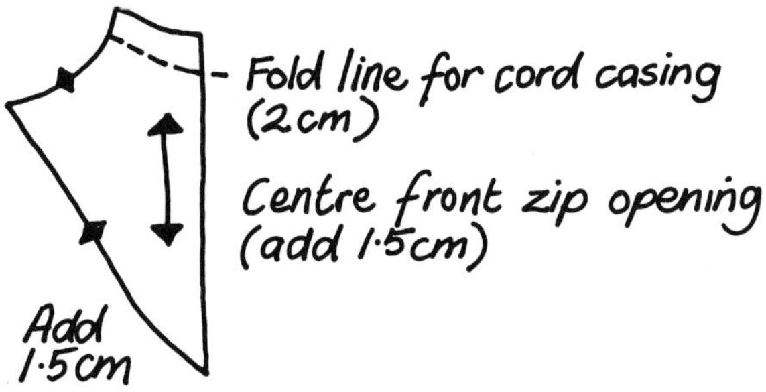

BACK COLLAR - cut 1 on fold

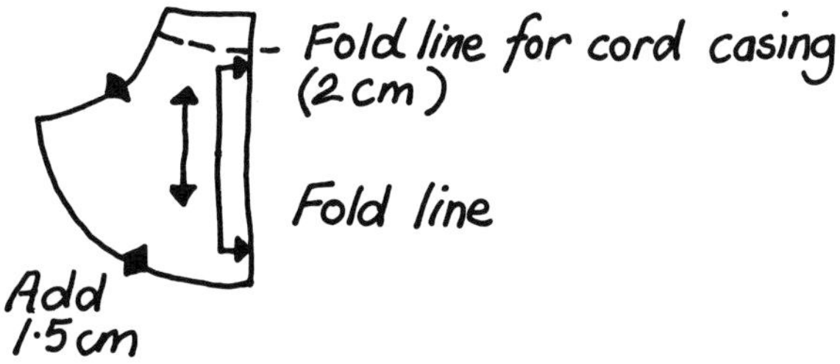

SLEEVE - cut 2

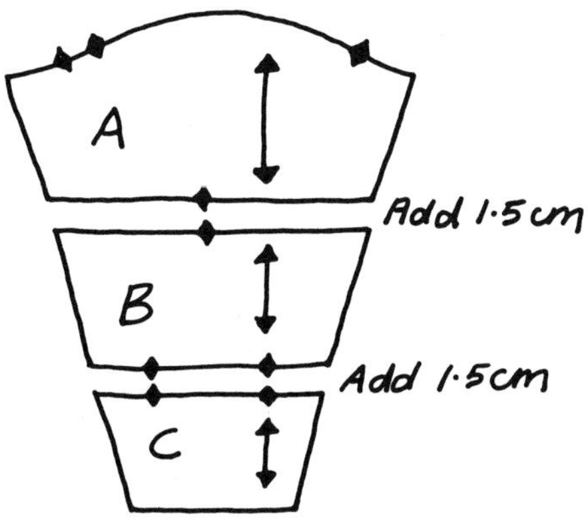

POCKET - cut 1 on fold

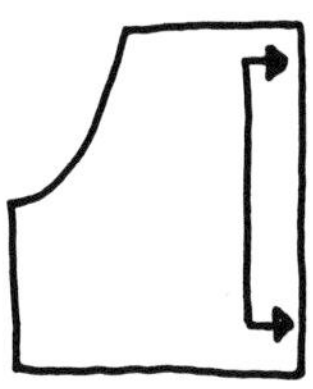

MINI POCKET - cut 2

FRONT - cut 1 on fold

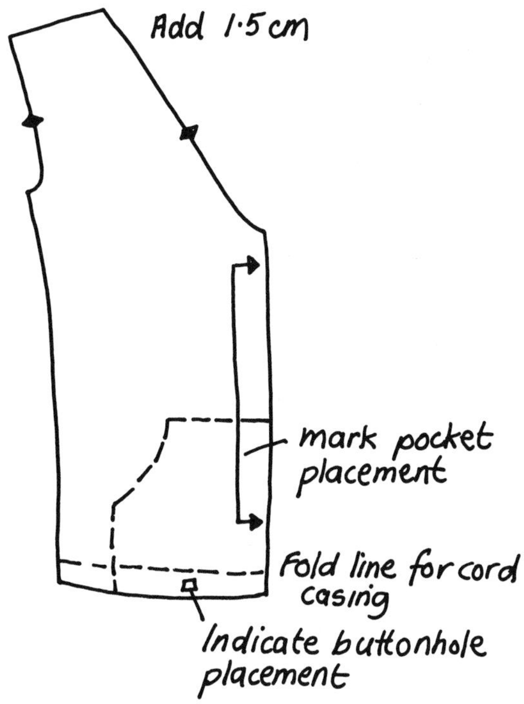

BACK - cut 1 on fold

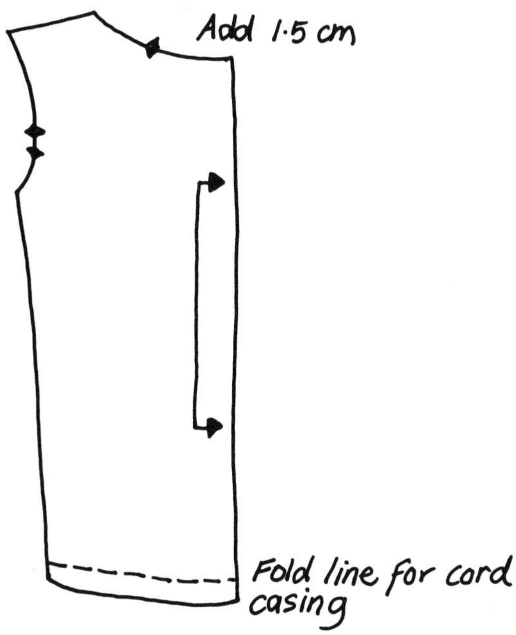

WINDCHEATER CONSTRUCTION (cont'd)

Now you have established the pattern pieces and tools, equipment and processes required to complete the garment you will need to prepare a schedule of operation (work plan).

For example

1 Carefully cut fabric to pattern and mark all notches and construction lines.

2. Insert a centred zip in front collar and mini pocket (see construction processes). Topstitch accordingly.

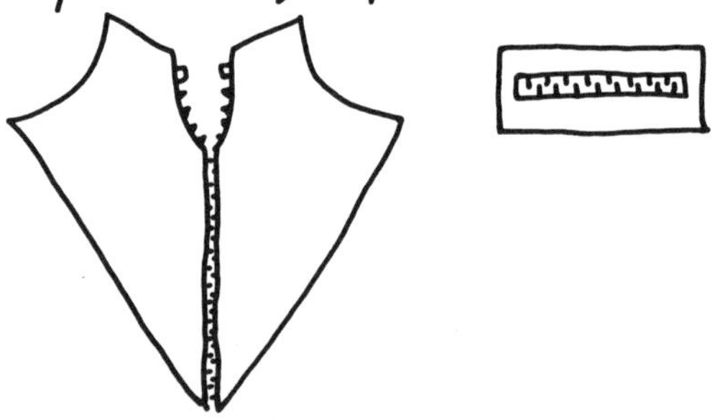

3. Turn under seam allowance on all edges of mini pocket. Tack into place on main pocket. Top stitch accordingly.
Follow remaining pocket directions according to pattern instructions
Sew button holes for bottom cord casing.

4. Join front collar to back collar at shoulder and neck seams (right sides facing).

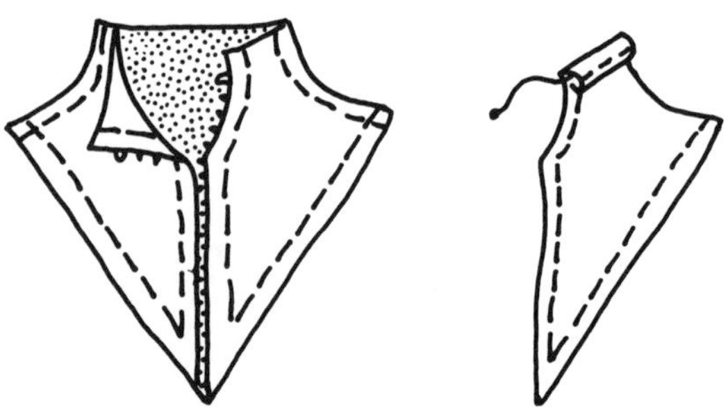

5. Turn down 0·5 cm around neck edge and fold down a further 1·5 cm. Stitch into place (similiar to bottom edge) for neck cord casing.

6. Stitch front to back at shoulder seam (right sides facing).

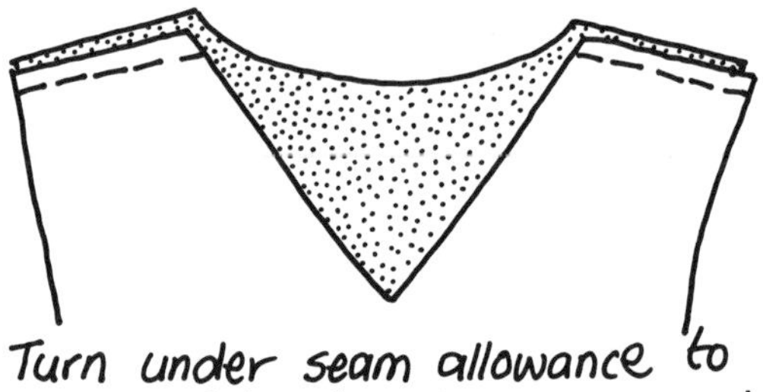

7. Turn under seam allowance to wrong side of collar edge and carefully tack into place onto main body. Carefully top stitch into place (remove tacking).

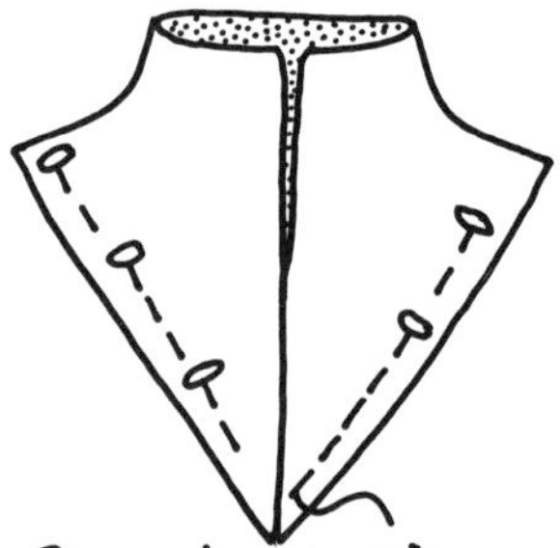

8. Sew sleeve pieces A, B and C together to make 2 full sleeves. Top stitch appropriately.

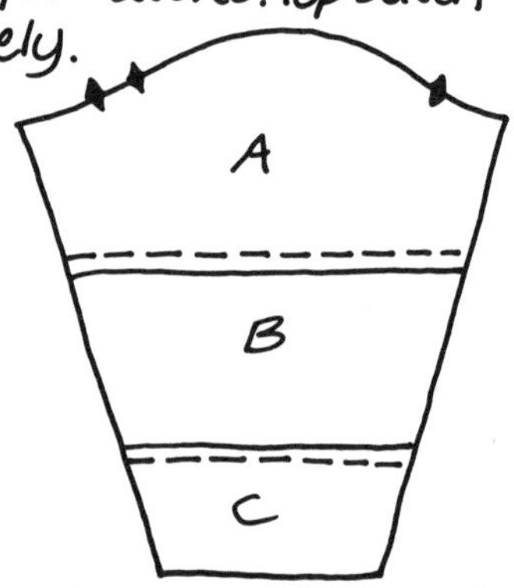

9. Attach sleeve to garment according to pattern directions.

10. Sew front to back underarm sleeves and side seams (right sides facing).

11. Turn up cord casing to wrong side and stitch into place.

12. Turn up sleeve hem and stitch into place.

Evaluate the finished product using the criteria you set out in your design plan.

CONGRATULATIONS on a job well done. You now have your own designer windcheater.

Other activities you may like to try could be:

- Track Pants—possibly add zips, net inserts, knee patches etc.
- Shorts, Skirts, T-Shirts and other matching sports and casual gear.
- Body Suits, Swimwear, Gym wear, Bicycle gear, anything that gives you skills using stretch fabric and elastic.
- Sports bags or other travel accessories.
- Decorator Cushions. Cushions can cover a wide variety of decorative skills. For example, piping, frills, appliqué, patchwork, pleating, embroidery and much, much more.
- Table Linen and matching accessories, like place mats and serviettes.
- Bedroom Furniture. This can include doona covers, pillow slips, valances, cushions, curtains—this industry has boomed in the past few years and there are lots of catalogues to give you some good starting ideas.
- No doubt you have more interesting ideas of your own and can add to this short list of things to make and explore.

STUDENT ACTIVITIES

1 Prepare some samples using the different decoration techniques outlined. Consider using a combination of these. For example, hand painting with beading and embroidery or silk screen on tie-dyed fabric.

2 Find the meaning of:
Motif
Tailored
Decoration
Notching
Construction
Catalogues
Contrast
Androgynous
Notions

CHOOSING THE RIGHT MATERIALS

Successful sewing or any application of textiles depends to a large extent on the right choice of fabrics for your particular purpose. In this chapter we will discuss some of the fabrics available, their characteristics and what they are most suitable for. Some tests have been suggested which will establish guidelines for judging the performance and suitability of different materials and how best to care for them. Where possible, testing should relate to your production. This chapter later explores the requirements of your report.

The focus of this chapter is:

- **guide to choosing fabrics**
- **fabric properties**
- **suggested materials testing and evaluation activities**
- **workshop testing activity report.**

GUIDE TO CHOOSING FABRICS

It is important to understand fabrics and their properties so we can choose them for the purpose to which they're most suitable. If you're going to the time, expense and trouble of designing and making clothes, you will rely on your expertise to make the right decision. We have all seen examples of garments and furnishings made from poorly chosen fabrics and the disastrous effect it can have. Hopefully, this section will give you some helpful tips on how to be 'consumer wise' and make the right choice each time.

As explained earlier, fabrics are made from many different types of fibres which may be used singly (as in 100% cotton or pure wool) or in a combination of two, three or more different fibres (such as wool-acrylic blend or cotton-polyester-viscose).

Apart from silk, all natural fibres are short and called staples (remember the wool story). Silk and some more recently developed synthetic fibres are already formed into long, continuous strands called filaments. The staples and filament lengths are then twisted into yarns which are used to make the fabric. The degree of twist in a yarn will determine the appearance and durability of the fabric. Woollen and worsted yarns are classic examples to compare (See Chapter 3). The tighter the twist generally means a smoother, stronger fabric. Refer back to Chapter 1 to refresh your memory on the different construction techniques of fabric.

Colour and pattern of fabric is largely a matter of personal choice and the trends and fashion influence of the season. The amount of money available can also determine your choice of fabric. Here are a few tips to assist you in choosing suitable fabric for a garment.

Name of fabric	Origin or fibre blend	General properties	Laundering instructions	Price	Available from
		e.g. drape durability strength texture absorbency			

Fig. 5.1 My personal fabric file

1. Drape the fabric either across your hips or over your shoulder, depending on what you intend to make. This will indicate whether the colour suits your skin tone and establish whether the fall and weight of the fabric is suitable for the garment you have in mind.
2. Swatches of fabrics are great—particularly if you're matching colours or patterns. Most stores are generally willing to give you a small sample.
3. Keep notes on fabrics you use and new fabrics you see.
4. Check if the fabric is likely to fray excessively. If it does, you are likely to encounter some inconvenience during construction.
5. Check if the pattern or weave has a 'one-way' design or a nap. Look at your pattern to see if extra fabric is required.
6. Carefully inspect the fabric for any flaws or marks when you buy it. Note the fibre composition as this will determine its cleaning requirements.

The most desirable properties of fabrics are:

- strength; for durability to cope with general wear and tear and constant laundering.
- texture; to ensure that fabrics are non-abrasive and non-irritating.
- elasticity; if the fabric has stretch (give), it will retain its shape and show less creases.
- dye/colour; most fabrics are dyed and it is important that fabrics retain their colour and don't fade dramatically.
- laundering; all fabrics must be cleaned. It is more convenient and cheaper if we can do this ourselves. Dry-cleaning and special care of some fabrics is usually time consuming and more expensive.
- absorbency; for general comfort, a fabric should be able to absorb a certain amount of body moisture (otherwise we would wear plastic).
- lustre; shiny, smooth fabrics collect less dirt than a rough, textured surface.
- price; for general everyday garments, fabrics should be available at a realistic price to keep them affordable. More expensive garments and fabrics are generally worn on special occasions.

Table 1.3 gives background knowledge of fibres and fabrics and will help you in your testing of materials. It will assist you in determining the most suitable fabric for a given purpose. Check the fabric you have in mind and see if it has the properties you are looking for. Perhaps you will find a more suitable alternative.

Commercial patterns usually make recommendations as to the most suitable fabrics for that style taking into account the drape of the garment, its purpose, whether it's suitable for stripes, plaids or checks, and other factors.

Figure 5.3 is an example of a commercial pattern and the information usually provided.

TESTING AND EVALUATING MATERIALS

The textile industry of Australia has recently released the new GarmentMark—an all Australian label which is the sign of quality. The specifications for manufacturers to be included in this particular scheme cover areas such as colour fastness, stretch recovery properties, and pilling of wool or polymer fabrics.

Industry is continually testing fabrics for their suitability for a given purpose. Materials technology places a great deal of emphasis on getting to know and understand materials we commonly use and those which are being modified or newly developed. The tests outlined here will keep you up to date on industry's latest innovations and give you a sound knowledge of fabric.

GarmentMark
All Australian

Fig. 5.2 The GarmentMark logo

4" (10cm) OF CROSSWISE FOLDED KNIT MUST STRETCH FROM HERE → TO HERE OR MORE →
4" (10cm) DE JERSEY PLIE SUR LE TRAVERS DOIVENT S'ETIRER D'ICI → A AU MOINS LA. →

5606

MISSES', MEN'S OR TEEN BOYS' TOPS AND PANTS OR SHORTS AND PERMANENT TRANSFER - *FOR STRETCH KNITS ONLY:* Very loose-fitting pullover tops have dropped shoulders; view A has neckband, sleevebands and lower band cut of ribbed knit; view B has a hood, front patch pocket and purchased cording for drawstring at lower edge; permanent transfer also included; loose-fitting pull-on pants with elastic at ankles or shorts - both have fold back casing and three rows of elastic at waist, back patch and side seam pockets.

BODY MEASUREMENTS

	Misses' or Men's/Teen Boys'					
SIZES	X-Small	Small	Medium	Large	X-Large	
Bust/Chest	30½-31½	32½-34	36-38	40-42	44-46	Ins.
Hip	32½-33½	34½-36	38-40	42-44	46-48	"
VIEW A - Top						
58/60"*	1⅝	1⅝	1⅝	1¾	1¾	Yds.
Contrast (Neckband, Sleevebands,and Lower band)						
18" Wide Tubular Ribbed Knit **or** 36" Wide Flat Ribbed Knit, ½ yd.						
VIEW B - Top						
58/60"*	2½	3	3	3	3	Yds.
Pants						
58/60"*	2⅜	2⅜	2⅜	2⅞	2⅞	Yds.
Shorts						
58/60"*	1⅛	1⅛	1½	1½	1½	Yds.
FINISHED GARMENT MEASUREMENTS						
Back length from normal neckline						
Top B						
Misses'						
Teen Boys'	29¼	29¾	30¼	30¾	31¼	Ins.
Men's	30¼	30¾	31¼	31¾	32¼	"
Back length from shoulder						
Top A						
Misses'						
Teen Boys'	27¼	27¾	28¼	28¾	29¼	"
Men's	28¼	28¾	29¼	29¾	30¼	"
Side length from natural waistline						
Pants						
Misses'						
Teen Boys'	40¼	40¾	41¼	41¾	42¼	"
Men's	42¼	42¾	43¼	43¾	44¼	"
Measurement at bustline/chestline						
Top A	43½	46	50	54	58	"
Top B	47	49½	53½	57½	61½	"
Measurement at hipline						
Top A	42½	45	49	53	57	"
Top B	47	49½	53½	57½	61½	"
Pants, Shorts	45	47½	51½	55½	59½	"

***With Nap** - Additional Fabric may be needed to match stripes or plaids.

SUGGESTED FABRICS: All Garments - *For Stretch Knits only* such as Sweatshirt Knits • Cotton Knits • Double Knits; **Top A** (Contrast Neckband, Sleevebands and Lower Band) - Tubular or Flat Ribbed Knit. **NOTE: Top B** - Wrong side of fabric will show when sleeves are rolled up. **NOTE: All Garments** - Not Suitable For Diagonals.

NOTIONS: Thread; **Top B** - 4½ Yds. of ⅛" Wide Cording; **Pants** - 4½ Yds. of ½" Wide Elastic; **Shorts** - 3¾ Yds. of ½" Wide Elastic.

16 PATTERN PIECES

PULLS AVEC DECALQUE PERMANENT ET PANTALON OU SHORT, POUR JEUNE FEMME, HOMME OU JEUNE HOMME - *EN JERSEY EXTENSIBLE SEULEMENT*: pulls très amples à emmanchure tombante; bande inférieure, bande d'encolure et bandes de manches en Jersey à côtes pour Modèle A, capuche, poche plaquée devant et cordon tout-fait coulissé dans la partie inférieure du Modèle B; décalque permanent également inclus. Pantalon souple avec élastique coulissé aux chevilles ou short - chaque modèle avec trois rangs d'élastique coulissé à la taille, poche plaquée au dos et poches dans les coutures de côtés.

MESURES

	Jeune Femme ou Homme/Jeune Homme					
TAILLES	Toute-petite	Petite	Moyenne	Grande	Très-grande	
Tour de poitrine	78-80	83-87	92-97	102-107	112-117	cm
Tour de hanches	83-85	88-92	97-102	107-112	117-122	cm
MODELE A - Pull						
150cm*	1.50	1.50	1.50	1.60	1.60	m
Parties contrastantes (bandes d'encolure, bande de manches et bande inférieure)						
Jersey à côte tubulaire de 46 cm ou non tubulaire de 91cm, 0.50m						
MODELE B - Pull						
150cm*	2.30	2.70	2.70	2.80	2.80	m
Pantalon						
150cm*	2.10	2.20	2.20	2.60	2.60	m
Short						
150cm*	1.00	1.00	1.30	1.40	1.40	m
MESURES DU VETEMENT FINI						
Longueur dos depuis encolure naturelle						
Pull B						
Jeune Femme						
Jeune Homme	74	76	77	78	79	cm
Homme	77	78	79	81	82	cm
Longueur dos depuis l'épaule						
Pull A						
Jeune Femme						
Jeune Homme	69	70	72	73	74	cm
Homme	72	73	74	76	77	cm
Longueur côté depuis taille naturelle						
Pantalon						
Jeune Femme						
Jeune Homme	102	104	105	106	107	cm
Homme	107	109	110	111	112	cm
Mesure à la poitrine						
Pull A	111	117	127	137	147	cm
Pull B	119	126	136	146	156	cm
Mesure aux hanches						
Pull A	108	114	124	135	145	cm
Pull B	119	126	136	146	156	cm
Pantalon, Short	114	121	131	141	151	cm

***A sens** - Il faut davantage de tissu pour raccorder les rayures ou les écossais.

TISSUS CONSEILLES: Tous vêtements - *En Jersey extensible seulement* tel que Jersey molletonné • Jersey de coton • Jersey double; **Pull A** - (bande d'encolure, bandes de manches et bande inférieure) - Jersey à côtes tubulaire ou non tubulaire. **N.B. Pull B** - L'envers du tissu se verra lorsque les manches seront retournées. **N.B. Tous vêtements** - Les diagonales ne conviennent pas.

MERCERIE: Fil; **Pull B** - 4.15m de cordon de 3mm; **Pantalon** - 4.15m d'élastique de 1.3cm; **Short** - 3.45m d'élastique de 1.3cm.

16 PIECES DE PATRON

Fig. 5.3 Back of a pattern envelope suggesting fabrics

Where possible, it is best to relate your tests to your work. For example, if you are making a woollen jacket you may like to compare wool with a polyester blend for water absorbency, fire resistance and crease recovery. We suggest you look at a local manufacturer and find out what tests they conduct to choose appropriate fabrics.

Depending on your production and the focus of the unit you are studying, you might like to test some related materials. You may decide to run a series of tests to determine the suitability of a fabric you aren't familiar with. Alternatively, fabrics are forever changing as modifications are made to existing fabrics. This provides you with an excellent opportunity to test the old with the new and ascertain which is the best for your purposes. As you will see, there are endless opportunities for choosing suitable tests and you will find that the tests are quite enjoyable. You will also learn some interesting details of the materials we use. It is a good idea to check your proposed tests with your teacher before you begin so she or he has the opportunity to approve your testing procedures. Discuss all aspects of the test and make sure that you understand the method clearly.

When the purpose and procedure of your test is clear, you can begin.

Don't worry that testing materials requires you to perform elaborate scientific procedures in a laboratory. The skills of selection, observation and evaluation will enable you to make confident decisions relating to materials and how best to use them. Writing an appropriate report is also an essential component of the process as it acts as a means of communicating your new found knowledge to someone else.

RECORDING AND REPORTING RESULTS

Below we have provided a sample format to document all the information relating to each test that you perform. This report should be written. You may like to support your results with diagrams or graphs where appropriate.

MATERIALS TESTING AND EVALUATION: WORKSHOP TESTING ACTIVITY REPORT

The following should be presented in written form under the following headings:

Materials tested

Give the names of the materials involved in the test. Describe the materials and identify their essential differences. Describe the specific characteristics being tested. Is this an appropriate test for these materials?

Purpose of the test

What are you attempting to prove? Comment on your predicted result. What criteria will you use to make judgements regarding the materials' suitability for a given situation? Why are these criteria suitable?

Description of the test

What method did you use for testing? Why did you use specific tools, equipment and machines? Briefly state your reasons for selecting your tools and equipment. Describe a possible alternative. Comment on the relative advantages and disadvantages of the alternatives. Why did you follow specific processes?

Results

Record all of your observations. Present your observations in an appropriate form and ensure that your results are relevant to the stated aim of the test. How accurate were your observations?

Conclusions

Comment on the effectiveness of the tests. Comment on the actual result versus your prediction. Were your procedures appropriate? How do your results compare with other information relating to the materials tested? Could you repeat this test?

Recommendations

Give recommendations about the appropriateness of the materials for a given situation.

Resources

Be sure to attach a list of all the references you used for your report.

STUDENT ACTIVITIES

The following section includes a range of testing activities you may like to use as a starting point.

1 Shrink Test
Select a variety of fabric samples, ensuring that they are from a variety of sources.
   a. Make sure that each piece of fabric is exactly the same size (10 cm square).
   b. Soak each piece of fabric in hot water for at least 10 minutes.
   c. Remove the samples and carefully squeeze the excess water and allow to dry. When all the samples are dry, iron and remeasure them.
   d. Record your results and complete a report following the report format outlined earlier.
   e. Keep the samples and add them to your report as part of your results.

2 Wrinkle Test
You may like to use the same fabrics you used in the shrink test.
   a. Hold each piece of fabric in your hand and crease it up. Hold it in this position for 15 seconds.
   b. Release the fabric and observe whether the sample is creased. Record your results: no creases visible, some creases visible, very creased.
   c. You may be able to rate the fabrics on their recovery rate.
   d. Include the fabric samples in your report as part of your results.

3 Design some tests to examine the effect of bleaches, dyes, detergents, fabric softeners and protective treatments (for example—Scotchgard) on a range of fabric samples.

4 Compare silk with polyester for its absorption, evaporation, flammability, laundering qualities and perhaps some microscopic testing. From your results, suggest some suitable applications for both fabrics and justify your answer.

5 Carry out the Burning Test. Use the report guide to prepare a suitable report.

6 Read the article 'Challenging Fibres'. As a class, discuss the stigma attached to synthetic fibres. Do you think they have had an effect on Australia's wool and cotton industries? Should there be policies to protect such valuable industries or should we keep promoting synthetic fibre development?

7 Make a list of the advantages and disadvantages of using Lycra. What special care do you need to take when sewing Lycra? What seams would you use? Justify your answer. Obtain some samples of fabric containing different quantities of Lycra and suggest possible applications for each.

8 Paddy Pallin manufacture modern outdoor clothing and have researched and developed fabrics to cope with extreme weather conditions. Read through their extensive list of fabrics and their properties. Contact them for more information. Read the philosophy behind Integral Clothing and the rigorous design process that produced these garments. Perhaps you could set yourself a similar design brief as an extension activity and incorporate your knowledge of these new innovative fabrics into your designs.

Burning Test – General Procedure

Adapted from an article by Carolyn Paulin of Victoria College, Rusden Campus, published in Echo Vol 31 No 1 1992.

Fibres, yarns and pieces of fabric can be used as specimens. If the product contains a combination or blend of fibres, test several threads from weft and warp to see if they have the same fibre content.

Care must be taken to prevent burning your fingers or inhaling smoke from the burning samples. Hold the samples with a pair of tweezers. Do not allow any molten material to drop onto your skin. Use a protective mat to catch burning material. Be sure all material is cold before disposing of it.

Test Procedure

1 ***Describe the material before testing***
What colour is it? What is its texture? Does it have an odour?

2 ***Approaching the flame***
Slowly move the fibre or yarn sample towards a small flame. Observe the reaction of the sample to the approaching flame. Does it ignite readily? Does it shrink away from the flame?

3 ***In the flame***
Place the sample directly in the flame to determine its burning characteristics. How does it burn? Does it burn quickly? Does it go out quickly? Does it burn with melting? Is any smoke produced? Is there any odour?

4 ***Remove from the flame***
Remove the sample from the flame and observe its behaviour. Is there an afterglow? Does it continue to burn? Does it continue to melt or smoke? Is there any odour?

5 ***The residue***
Allow the sample to cool and check the characteristics of the residue. How does it smell? Does it resemble burnt paper? Does it resemble burnt hair? Can you crush the residue between your fingers?

Fig. 5.4 The Burning Test

BURNING TEST

DIRECTIONS:
Unravel and test several yarns from warp and weft of fabric to see if they have the same fibre content.
Hold each yarn horizontally and feed slowly through edge of flame of taper. Repeat for double check.

FIBRES	WHEN APPROACHING FLAME	WHEN IN FLAME	AFTER REMOVAL FROM FLAME	ODOUR	RESIDUE
Cellulose Cotton Flax Rayon	Does not fuse or shrink from flame. Ignites readily.	Burns rapidly	Afterglow	Burning paper	Grey feathery smooth edge
Protein Silk Wool	Fuses and curls away from flame	Burns slowly	Self-extinguishing	Burning hair	Crushable black ash
Acetate	Fuses away from flame	Burns with melting	Continues to burn and melt		Brittle black hard bead
Acrylic	Fuses away from flame	Burns with melting	Continues to burn and melt		Brittle black hard bead
Nylon	Fuses and shrinks away from flame	Burns slowly with melting	Usually self-extinguishing	Celery-like	Hard grey bead
Polyester	Fuses and shrinks away from flame	Burns slowly with melting, black smoke	Usually self-extinguishing	Sweetish odour	Hard black bead
Spandex	Fuses but does not shrink from flame	Burns with melting	Continues to burn with melting		Soft black ash

'CHALLENGING FIBRES'

Pure versus synthetic. Is one really better than the other?

Now that the stigma associated with man-made fibres has been laid to rest, JANICE BREEN looks at what's new on the fabric front.

As if they did not have enough problems recently, fashion houses have also had a hard time luring recession-weary, neo-conservative, cash-strapped customers back to the ready-to-wear racks.

Seduction by design is not easy when what your market craves most in her wardrobe is practical flexibility, comparative blandness and ultimate longevity. If it goes with a dozen other pieces she already has and it could still be worn in five years, the theory goes, it is not a bad buy.

Designer fashion stands to lose most as its market's preference slips downward into a less emotive and less expensive middle range. But in Europe and America and increasingly here, designers are checking the slide by using tricky fabrics to grab their — at best finicky, at worst listless — customers' attention. In the fashion houses that can afford the most seductive cloths, business is now truly a matter of seduction and "the look" is as much about garments' palpable, evocative qualities as it is about aesthetics.

"Buying (fashion) has become a passionate, tactile affair," says Melbourne designer John Cavill, "a matter, partly, of having a feel."

Fabric as a seducer is not a new concept to Mr Cavill. He has criss-crossed Europe fossicking for its finest fabrics more times than he can remember. He talks about them as if they have little personalities. "I've used this stretch (fabric) for summer. It's a mix of acetate, viscose and Lycra but that's not what is important. It has depth and intensity. It's grippy for a slimmer style. It's got bounce. It skims . . . " He says the old stigma associated with synthetic fabrics is all but dead, although garments tagged "pure" still have pull. "Reading labels just isn't part of the buying process anymore. It's more, 'how fabulous is *that* . . . ?' Whether it is synthetic doesn't matter."

Little wonder. Some remarkable cloths, laid to rest decades ago by pure wool, cotton, silk and linen snobbery have resurfaced fresh and lovelier than ever. Long strides in new synthetic fabric technology have also borne cloths that move, roll, flop, tumble, slink, flap, then return almost flawlessly to their pre-crease state, wash like a dream and in many cases to not even need pressing. Colors are pure and resist fading, textured surface patterns are as rich as brocade or subtle as jacquard silk, and although the question — "Is it cheaper than pure?" — is a bit like asking how long is a piece of string, the answer is usually, "Usually".

One of the technology's finest fruits is the remarkable all-synthetic stuff called microfibre. The thinnest microfibre fabric can flutter on a kitten's breath and has the sensuous, clammy softness of sanded silk.

"In fact, you can hardly tell the difference (from silk)," Wenzel fabrics' Naomi Menahem says. Her firm can supply three different microfibre weights from shirting to suiting and the orders are swelling with every season. "Silk is delicate, it does mark easily, and tends to fade with washing. Microfibre is easy care; it breathes, feels and drapes like silk, without the hassles."

The sturdy structure of microfibre-fabrics is also a boon to designers bent on reinventing the wheel. Surprising applications of incredibly beautiful fabrics in unexpected garments is all part of the new saleable seduction.

Light, strong, toasty-warm and water repellant microfibre jackets and coats wowed them in recent winters from New York to Milan where the old symbol of status and common-sense, the furt coat, is now regarded as an ugly no-no. The microfibre alternatives were hailed as smarter by far, heaven to look at and a dream to wear. One version could be converted to a swishy city duster by detaching its inner shell of (rabbit) fur or wool. Others were constructed from an unbelievably soft, fleshy sandwich of synthetic filling between two silky microfibre skins.

Journalists' search for superlatives worthy of microfibre (languid, cheeky, liquid, lazy to name a few) is reminiscent of the fracas for Du Pont's introduction of Lycra two decades ago, and later, the Australian Wool Corporations's first release of Cool Wools.

Ted Steel, manager-Lycra at Du Pont, was there at the beginning when Lycra replaced rubber in girdles and among its myriad other applications, spawned hosiery that could be simultaneously sheer as gossamer, comfortable and run-resistant. A miraculous gift from the techno-gods, Lycra also fuelled the fashion bodywear revolution. Lycra was *the* bodywear revolution, and still is.

These days Mr Steel is monitoring the amazing stretchy yarn's integration in everything from shoes to men's knickers. Second-skin fit is not Lycra's only forte. It can, and is being incorporated in a huge range of fabrics and ultimately, a variety of fashion outerwear because of the tactile quality, the cleaner, floppier look, the modicom of "give" and the pre-crease "memory" that it imparts on its host fabric.

"Just two or three per cent of Lycra in knits or wovens can change the drape and handle," Mr Steele says. "In slacks a little (Lycra) means they will give when you sit down. They also hang better and return to shape when you stand up."

It is this "intelligent" aspect of synthetic yarns which another Melbourne designer, Rae Ganim, says is pecking away at her middle market's hitherto-staunch penchant for pure fabrics. Ms Ganim says her middle-market customers approach fashion from a different direction to the markets above and below: more considerate, slower to change and resistant to fads. She observes that further up market, customers are more finely tuned to fashion's emotive, seductive qualities, some of which are realised in exotic, synthetic fabrics. This has eroded that market's traditional snobbery about pure fibres. Further down market too, customers of budget brands are aware of the economy of cheaper synthetic fashions. Erosion of that market's preference for pure fabrics is a practical choice.

Ms Ganim's middle market is composed of swinging voters. Her label is a sort of half-way fashion house which, according to her husband and partner, Anthony Ganim, attracts customers who are "too smart to be snobby" but at the same time, "are tuned in to the simplicity and honesty of pure fabrics".

"I think a lot of our customers prefer to buy pure fabrics," Ms Ganim says. "But people are practical, too. A pure cotton T-shirt can look like something the cat dragged in after a few washes. With 20 per cent synthetic content it can be easier to wash, easier to iron,the shape keeps . . . "

A touch of Lycra can also work wonders, she says, without compromising the loyalties of those pure fabric loyalists too much. "Even in a big, loose jacket, the feeling is softer and when you bend your arm, it bends with you. Straight skirts with a bit of Lycra also make good sense."

The Rae Ganim label constantly rationalises the good sense of synthetics with the good feeling of pure fabrics on the skin. It still has a broad reputation for purity, with a common-sense bent to impurity.

Ms Ganim optimistically predicts, however, that pure fabrics equal in all things to their synthetic peers will soon evolve. When that happens - "and there are some great pure cotton and wool fabrics here already" - the pure will again become her prime emotive and practical choice. "Because they feel good and ultimately, they are better for the environment and the '90s are all about feelings and the environment."

Ms Helen Rowe, group marketing manager of the new Australian Wool Corporation, could not agree more. She says the corporation's current advertising campaign featuring genuine wool growers explaining in slow, careful strine that wool is natural and good for us, our kids and our economy, is in harmony with "a global feeling, a new perception of reality".

This renewed reverence for honesty and simplicity blossomed in the bewildering vacuum left by the frantic, fastbuck '80s. Even the naming of wool's recent fashion palettes reflects this new, nicer '90s attitude in shades dubbed ochre, earth and vegetable tones.

The corporation is pushing to affirm wool's status as a fashion staple, a stalwart in a field of Johnny-come-latelies. "Synthetics come and go," Ms Rowe says. "Wool remains and it continues to be innovative. It's the fibre people would rather have, the fibre they would rather wear."

From The Age, Wednesday 13 May 1992.

Fig. 5.5 'Challenging Fibres'

INTEGRAL CLOTHING

SYNERGY (si-ner-ji) The action of two or more substances to achieve an effect greater than the sum of their individual effects.

Integral Clothing is a system of clothing designed to work in harmony with all other elements. For instance a Techtop collar opens indentically to a Slipstream to give maximum ventilation, a Spark fits neatly under a Voyager hood, a medium Vortex is cut to allow a medium Flash underneath.

There are many more subtle design facets which are appreciated with wearing. No garments, legwear or tops, have zips at the hipbelt level that could become hard pressure points. Every product with pockets always has a pocket accessible even when wearing a hipbelt, climbing harness, etc.

All but one of our tops has seamless shoulders to reduce rubbing and wear from pack straps.

Select Integral Clothing as high performance everyday wear or as functional backcountry "equipment" confident that every garment is as comfortable as the next.

AQUANORT is a 3 layer waterproof/breathable (WP/B) fabric. A microporous Polyurethane (PU) coating which is protected by a tough nylon outer and wicking inner of nylon tricot mesh. More WP/B than Milair but not as much as Gore-Tex.

CORDURA is a brand of rugged texturised nylon fibres woven into fabrics used mainly for packs and luggage. Also used on clothing for reinforcement of wear areas.

DRYTECH is a special knit of soft polyester fibres which transfer moisture away from the skin. The outer surface is a fine knit with narrow passages that rapidly draw perspiration from the open knit against the skin. Works like a one way valve keeping the wearer dry and comfortable.

EXODUS is an intimate blend of soft cotton (28%) and microbfibre polyester (72%). Its lightweight and minuscule pack size belie the fact it is four times more abrasion resistant than denim relative to its weight. Very windproof and breathable. SPF of 51.

FLEECE is a generic name given to fluffy polyester fabrics which insulate like wool but dry very fast, don't shrink and aren't itchy. Most are made in the States by Malden Mills which have recently changed their names. Polartec 100 (Polartek), Polartec 200 (Polarlite, Paddy Pallin's Pluslite), and Polartec 300 (Polarplus) are the new names.

GORE-TEX is the best available WP/B fabric. As a three layer it is a microporous membrane laminated between sturdy outers and wicking tricot mesh inners. Under most conditions moisture vapour (perspiration) can escape through the fabric while water molecules (rain) cannot penetrate. Scope Gore-Tex is a version used exclusively by Paddy Pallin for its great strength coupled with a supple and quiet feel.

INTERCEPT is a special three ply construction of performance fabrics. Exodus provides an outer shell of wind/showerproofing. Inside soft polyester micro-fibre Intercept wadding insulates without adding bulk behind a supple lining of Pertex.

LYCRA is a brand of elastomeric fibre. Fabrics with lycra in the blend are very stretchy with excellent recovery. However too much lycra and your garment may pill or be slow drying.

MILAIR is usually a 2 layer waterproof/breathable fabric. A non-poromeric PU coating on a durable nylon shell. Unlike micro-porous coatings Milair passes moisture through the fabric via molecular transfer. Works a little like gas escaping through a watertight plastic soft drink bottle to leave your lemonade flat. A budget WP/B.

PERTEX is a new technology in featherweight nylons. A very fine yarn is woven into a very dense weave which is warm-to-touch and moisture transferring. Uses other than for windshirts and garment linings include sleeping bags.

POLYPRO is short for polypropylene. A fibre used for underwear that is lightweight and tough. Polypro is waterhating so perspiration does not cling to the fabric but passes straight through into outer garments. This keeps the wearers skin dry. Avoid tumble drying and ironing.

Fig. 5.6 Paddy Pallin's Integral Clothing

DESIGN

We are all designers. Each day, most of us will design in some way. In this chapter we will look at designing and what to consider when we design. We will go through the first three steps of designing: the brief, the ideas, and refining.

Chapter 5 looked at the final product and the things to consider when designing. In any area of technology, irrespective of the materials used, design is a fundamental element.

The focus of this chapter will be:

- **the process of designing**
- **the Design Brief**
- **the Design Plan**
- **idea generation and design considerations, such as function, aesthetics, and ergonomics**
- **refining the preferred option.**

Designing has numerous definitions. Perhaps one of the better definitions of design is 'the process of finding the best option or solution in a particular set of circumstances'. To begin with we need to look at the different steps of designing.

Designing has many parts. In any project we attempt to work through the process shown in Figure 6.1. However, we rarely manage to do so in a logical manner. We are likely to have to design, trial, redesign and retrial many times. This means that we are likely to reach a final solution that suits the needs of the problem.

We assume that designing or redesigning occurs when a product needs improvement (although sometimes this is questionable, particularly in the fashion industry where the methods used in marketing sell the product rather than any design considerations). We should not need to redesign a product that works efficiently and looks effective. When designing, we often need to compromise in one area in order to accommodate another and therefore achieve a balance. For example, a shirt may look best made from silk, but it would be too expensive. Therefore, rayon, a cheaper fabric which looks similar to silk, could be used. Good design is being able to achieve a balance between the various elements of design. In textiles for instance, appearance is often the major consideration with less thought given to cost and function.

DESIGN BRIEF

A design brief is a short, clear statement describing the need or problem. It is essential that the brief is outlined prior to any work being done, as we must have clear guidelines before we can begin any project. If you are working for a client he or she should carefully check it before you commence designing. A design brief is often given to you by a teacher or client; it is up to them to make sure that is clearly set out. However, you may be required to write your own brief given a few guidelines. The main areas which need to be addressed in a brief are:

- stating the problem or project
- stating the function of the product
- special considerations and limitations.

Always keep it simple and clear as you will need to refer back to it regularly.

There can be a number of types of design briefs. For the purpose of this book we will use the following:

- Student initiated
- Teacher initiated
- Client initiated.

When you have a client, you have to design to needs which are outlined by someone other than yourself. This is by no means easy, but it is certainly challenging. There are a variety of skills that you will need, such as:

- How to select a client and how much do you need to know about them?
- How often do you need to consult with the client?
- How do you conduct your first interview with the client?
- What sorts of things do you need to record?
- Which decisions are made by you and which are made by the client?

Here are two examples of design briefs:

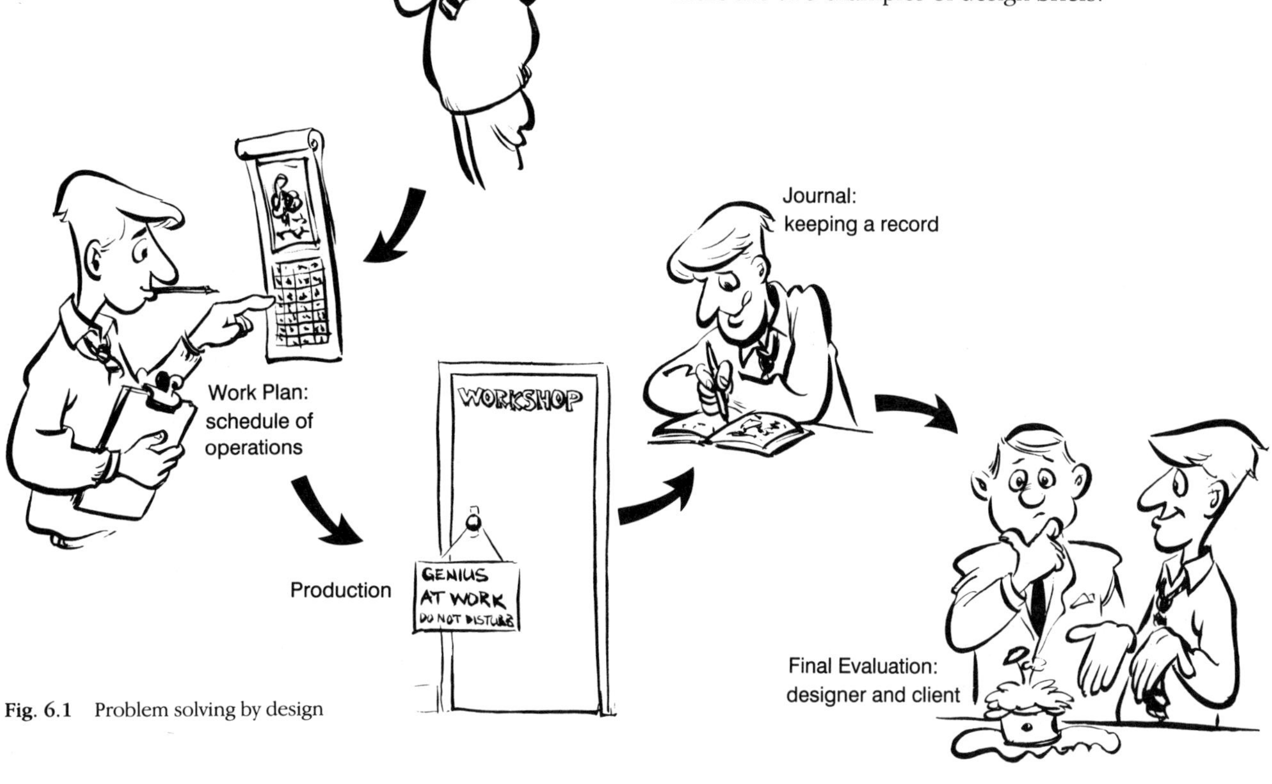

Fig. 6.1 Problem solving by design

TEACHER INITIATED

Students at your school are forever complaining about hard chairs. Your brief is to design something that will make them more comfortable. Considerations and limitations include cost, the ability to be mass produced, hard-wearing, and stackable.

CLIENT INITIATED

You have been asked to design an outfit for your 20 year old sister who is attending an afternoon wedding in Melbourne in June. The considerations include the climate; that the outfit needs to worn more than once; and cost.

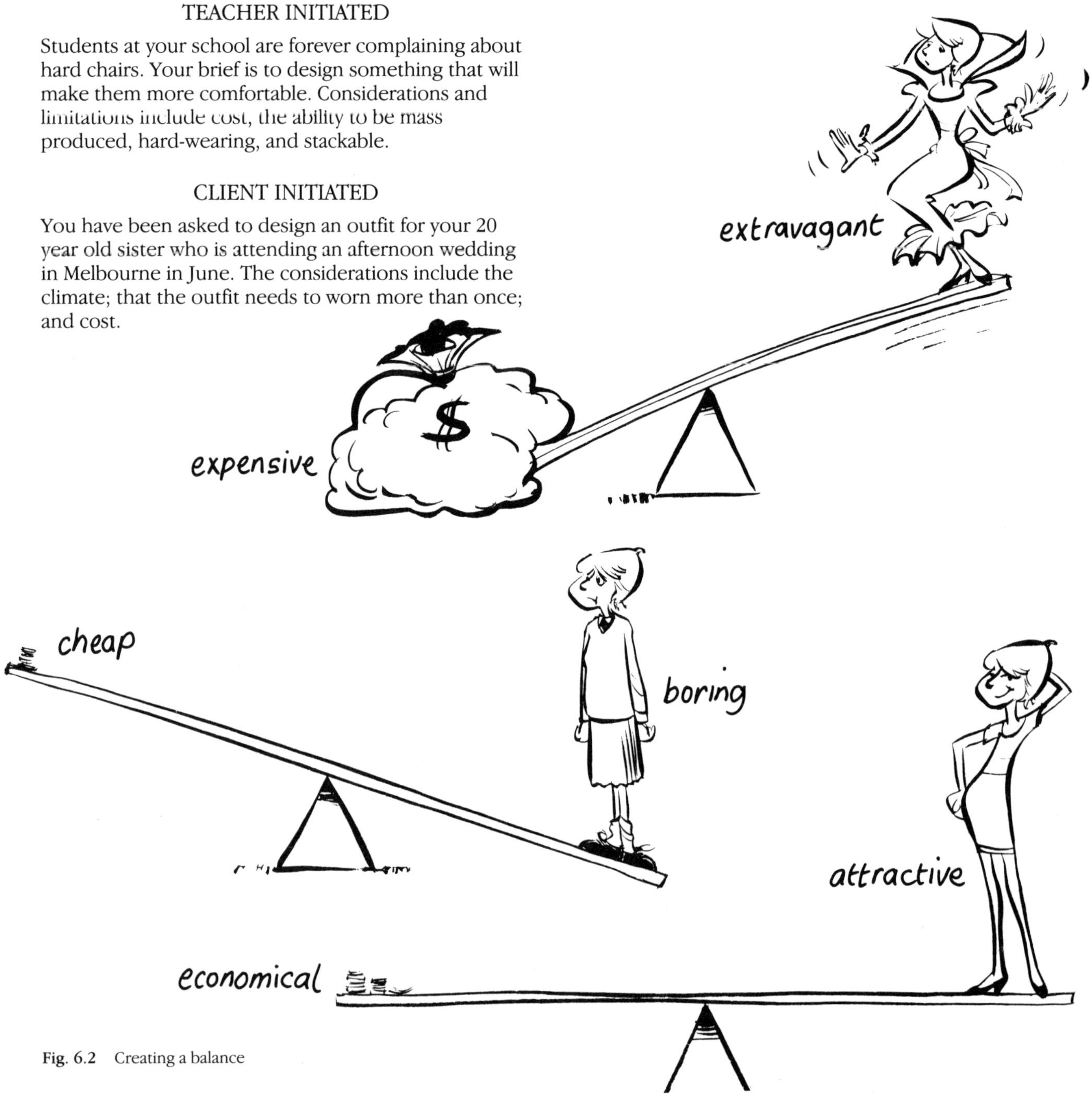

Fig. 6.2 Creating a balance

DESIGN PLAN

When designing, you may be required to produce a Design Plan, sometimes also called a sequence of events, production plan or a work plan. The Design Plan is simply a plan of your project and the steps you will follow. A Design Plan has two main aims:

- to provide a timeline for the planned project
- to provide a breakdown of each task within the project.

When we are planning, particularly if it is a large project, it is important to be able to break the work into smaller, achievable tasks. A plan may not be followed exactly, because we are constantly evaluating and reassessing tasks. However, it is a starting point. It is advisable and extremely useful to have a journal or log of the project and your progression as part of your plan.

IDEA GENERATION AND DESIGN CONSIDERATIONS

It is essential to explore different ideas and directions to ensure that we achieve the best result. This may not be (and often isn't) our original solution. At this stage there are many aspects of design which need to be looked at: function, aesthetics, ergonomics. We will discuss each of these areas briefly.

First we need to look at methods of investigating these aspects. It is important to investigate a product before designing or redesigning. A good starting point is brainstorming.

BRAINSTORMING

Brainstorming is about ideas. It is a method of writing down and thinking about lots of ideas relating to the

Task	Details	Date for Completion
Formulating Design Brief	consulting the client outlining the problem defining considerations and limitations	
Design Plan	planning time allowed allocating time to tasks formulating criteria for evaluation	
Trial Solutions	record and investigate sketch possible solutions make mock-ups	
Preferred Solution	selecting the preferred solution client consultation working drawings and plans presentation drawings	
Work Plan	clearly outline the plan for production	
Production	organising the pattern cutting fabric sewing main seams neatening seams decoration	
Journal	regularly updating the processes followed	
Evaluation	evaluating the final product self (designer) evaluation client evaluation user evaluation	

Fig. 6.3 A Design Plan

problem or project. It is a good idea to jot down ideas and words that come up while you are brainstorming as they can be useful in later stages of designing. Often brainstorming is done as a group or class, as the more brains there are, the more ideas there are!
Brainstorming is usually done in one of two ways:

1 through making comparisons. We do this by relating a problem or product to something we already know and work from there.
2 inquiring. We may not know much or, in fact, anything about the production or problems. We inquire—through people, books, resources of any sort—and use this as a starting point.

Now that you have a range of ideas, you can examine them more thoroughly. In other words, do some research.

FUNCTION

Function is the purpose for which a product was designed. There are two ways of analysing a product's function.

1 What is the product intended for?
2 What other functions does it have, even though they aren't its intended use. A chair, for example, is designed to be sat on—that is its function. But how many other uses are there for a chair? To stand on, to pile books on, to hang clothes over; a chair has many other functions!

When designing a product it is useful to complete a function analysis. This is not a difficult task; rather, it helps our designing and allows us to find out more about the intended purpose of the product enabling us to explore more ideas. A function analysis is a list of questions related specifically to the function of the product.

FUNCTION ANALYSIS

What is it?
What is its main purpose?
Is it easy to see what its function is?
What other things can it do or be used for?
How does it do it?
Identify each part of the product. What does each part do?
Are there any parts that may wear?
What care instructions are there?
Are any needed?
Is it safe? Are there any features which may not be so?
How long is it designed to last? Will it last this long?
Are the selected materials appropriate?
Who might use it?
What is the historical development of the product?
What alternatives could be used?
Are there any unnecessary features?
Is it good value for money?

You may not wish to use all these questions for all products. There may be questions of your own you can add.

Fig. 6.4 Brainstorming cushions

ERGONOMICS

Ergonomics is the study of the relationship between people and their working environment. When we design a product we are aiming to fulfil a need. Part of doing this is ensuring that the product is useable in its intended environment. We want the product to be comfortable, safe and efficient.

Ergonomics is more applicable to products such as home furnishings than to clothing.

Another word which is often mentioned when designing is anthropometrics, the measurement of the size and proportions of the human body. Anthropometrics is, of course, quite relevant when we are designing clothing.

In ergonomics we can also have a checklist. Things such as the size of buttons and clasps must be considered and, in furnishings, things such as size and height.

AESTHETICS

Aesthetics is the study of the nature of beauty. Aesthetics is related to visual aspects of the product meaning those areas or things we see. The appearance of the product plays a major role in marketing. This is particularly true of textile design and the fashion industry.

Often we look at products and make judgements about them which really relate to whether we like or dislike their appearance. However, we must be extremely careful when designing as appearance is very personal. To be able to objectively judge the aesthetics of a product it can be easier to break it into the elements of design: line, shape, form, pattern, texture, colour.

There are two ways to assess the aesthetics of a product:

- an aesthetics analysis
- looking at each of the elements of design.

One or both of the above can be used.

AESTHETIC ANALYSIS

All or some the following questions can be asked:
Is it well made? Why? Why not?
Is it a simple design?
Does the design suit the material used?
How much decoration is on it? Is the decoration suitable?
What colour is it? Does the colour suit the intended purpose?
Does it drape well?
Will it show dirt easily? Will it be easy to clean?
Will it crease easily?
Does its design appear balanced?
Are there any areas which don't blend in with the design?

Although we must admit that appearance is a very personal thing, we should still be able to assess why we like or don't like something.

DESIGN ELEMENTS

LINE

Line is a basic design element used to create style, shape, pattern, form and even texture. There are many different types of lines and these can have varying effects and meanings as they create different feelings. Lines differ enormously according to character, direction, dimension and the relationship between the lines. Some more common lines are:

- Vertical Lines; appear to add height, and make something appear more narrow. They lead our eyes in an up-down direction.
- Horizontal Lines; appear to shorten a shape or a figure. They emphasise width and imply solidity.
- Diagonal Lines; take on the feel of horizontal or vertical lines depending on their slant. They can add a sense of movement and instability.
- Curved Lines; create a softened effect. Their effect can appear greater on a curved body or in the drape of a curtain.

SHAPE AND FORM

Shape is made up of a series of lines or joined lines. The difference between shape and form can be said to be that shape is two dimensional and form is three dimensional. Different shapes create different feelings.

- squares and rectangles add stability
- circles give a feeling of movement.

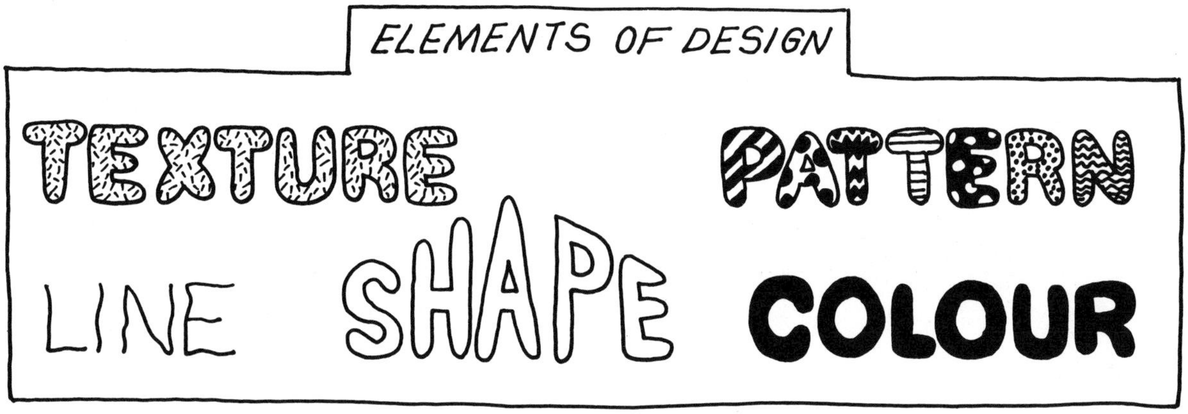

Fig. 6.5 Elements of design

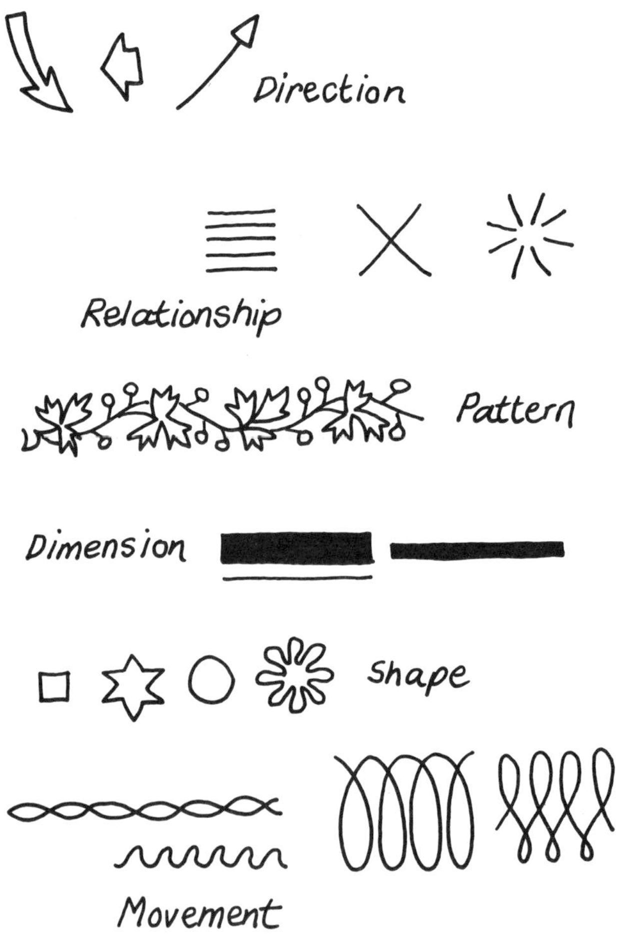

Fig. 6.6 Different properties of lines

Shape can be organic or natural (leaves for example) or geometric (precise). In clothing, we can look at shape in two ways. The overall shape of the garment or the shapes in fabric design.

COLOUR AND TONE

Colour is a major factor to consider in aesthetics. Perhaps it is the most obvious element and we react to it very strongly. We associate certain colours with things and colour is known to affect us psychologically. Firstly we will look at some of the colour terminology.

PRIMARY COLOURS

Red, yellow, and blue are colours which we cannot make by mixing other colours. We need the pigment of these to enable us to mix other colours.

SECONDARY COLOURS

Orange, green, and violet are made by mixing two primary colours together:

- Red and yellow gives orange
- Red and blue makes violet
- Blue and yellow makes green.

Fig. 6.7 Using lines

Fig. 6.8 Using shapes

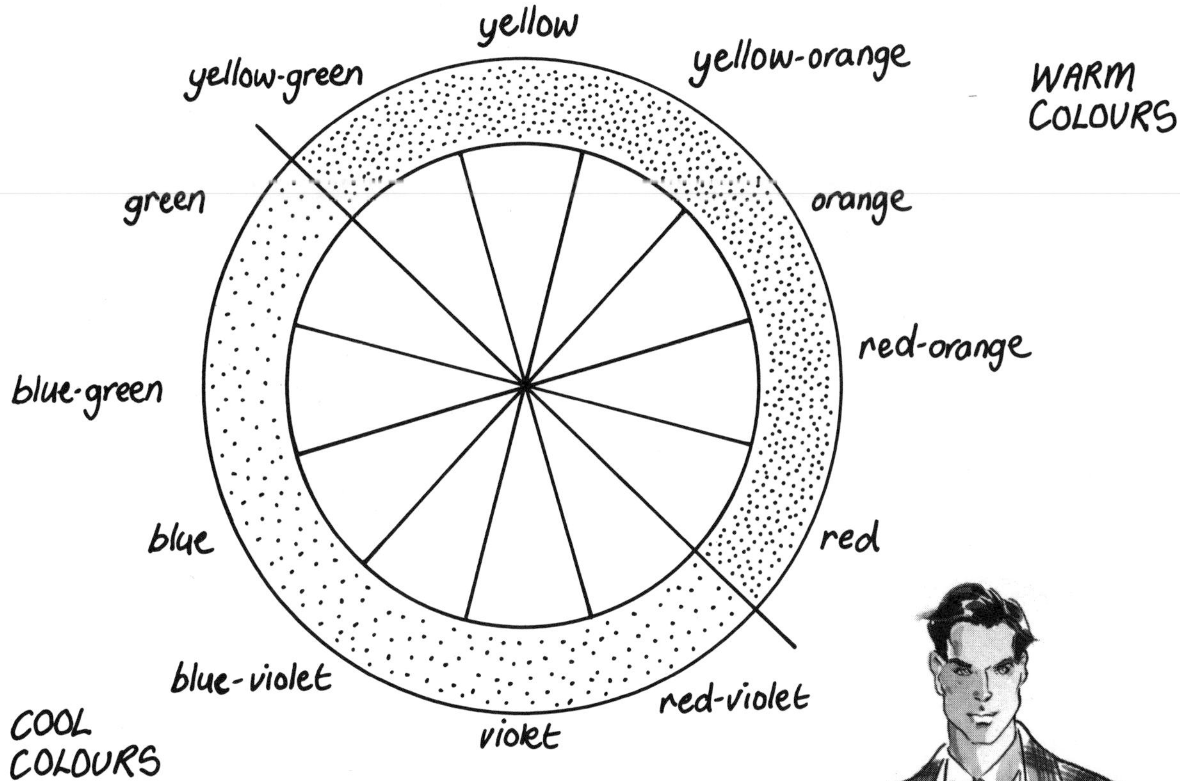

Fig. 6.9 The Colour wheel — use your colour pencils to colour it in

Extended Secondary Colours are those in between a primary and secondary, like red-orange and yellow-orange. These are also often called tertiary colours.

Different combinations of colours will have quite varying effects. For instance, combining colours next to each other on the colour wheel gives quite a different overall effect to combining those which are opposite. There are thousands of combinations you can use and you need to be careful that those you select suit the design of your garment or product.

An added factor of colour is tone. Black and white fall into this category. We refer to both as tones rather than colours. Black shades and white tints: if we had a blackish blue (or dark blue) we would call it a shade of blue and if we had a light blue, we would call it a tint of blue. Black and white can be used extremely effectively in three ways:

1 alone
2 with pure colours, for example, black, white, red and blue
3 mixed in with pure colours to make another colour, such as light blue and dark blue or light blue and light red (pink).

Don't forget grey. It is simply a tone.

PATTERNS

Patterns are a combination of lines or shapes. They can be simple or complex, depending on the amount of detail used. Complex patterns are more detailed, but not necessarily better because of this. Patterns can also be regular or irregular. The intended use or application of the product is the major influence on the effect of the pattern.

Fig. 6.10 Using tone

Fig. 6.11 Using pattern

Fig. 6.12 Using texture

TEXTURE

Texture is the way something feels, the surface of something. We have two types of texture:

- real, the actual feel of the surface and
- simulated or tactile, the appearance or look of the texture.

Texture can have a huge effect on the appearance of a product or garment. It must therefore be carefully considered in any design. For instance, a shiny surface will create a different effect to a fluffy surface or a harsh surface. In textiles, the feel of a material will determine how we can use it: a rough surface will not be used next to the skin, a slippery surface is inappropriate for a dining room chair. We must carefully consider the suitability of a fabric's texture before using it.

Once we have thoroughly investigated an area, we will have gained much background knowledge and be able to draw up a series of options. We will often gain ideas while investigating. Make sure you always put it down on paper! Draw it! Your drawings can be rough thumbnail sketches. It is often good to use annotated drawings; that is, drawings which have explanatory notes.

REFINING YOUR PREFERRED SOLUTION

Once we have a basic idea we need to refine it to get our preferred solution or final idea. To do this, select the most suitable idea and then consider the different elements of design—line, shape, colour, pattern, texture—and the material. It is often worth drawing the idea and copying it ten times to try out the changes. This is also a time when we must consider details such as decoration.

PRESENTING THE PREFERRED SOLUTION

When you have selected your final idea and refined it so that you have a clear idea of it, you need to be able to draw it. Drawing is done for two reasons:

- to sell your product indicating the 'look' of it, including shape, style and materials used
- to show how it is made.

When you draw to 'sell' your product, you need to make it look realistic and appealing. This is often done through a technique called rendering. Rendering makes a drawing appear three dimensional and shows tone and texture. Different areas of technology need to use different types of drawings to show their products and ideas clearly. In textiles we may need to draw a piece of furniture to show the textile covering or we may need to draw a garment. The two areas will require different types of drawing. The furniture will require you to complete a pictorial drawing, whereas the main emphasis in clothing will be on the figure. To complete a drawing of a figure, you will need to understand some basic rules about proportion. If you need to complete a series of garment drawings, complete one figure drawing and photocopy it several times. You can then 'dress' each figure as you wish.

Rendering can be done in many different ways and using different media. The materials you choose will depend on the effect you are after. You can use grey lead pencil, charcoal, crayons, pastels, aquarelle, markers, ink, paint, coloured pencils, airbrushing, watercolours, washes or finepoint markers.

Fig. 6.13 Rendering to make garments and objects appear realistic

WORKING DRAWINGS

Working drawings assist in the making of a product and therefore need to show details such as shape and size. If we were completing working drawings for something like a canvas chair, we would most likely use orthogonal drawings which show different views of the product. Sometimes this may also be accompanied by a sectional drawing.

In clothing, we are most interested on the cut and shape of the garment and each piece within the garment. The type of drawing we use needs to outline these factors. This is illustrated in Chapter 4.

Having completed all these tasks, you should have a clear idea of how to make your product. Chapter 4 deals with some of the issues relating to the making (or realisation) of your project. Below are some student activities which will help you.

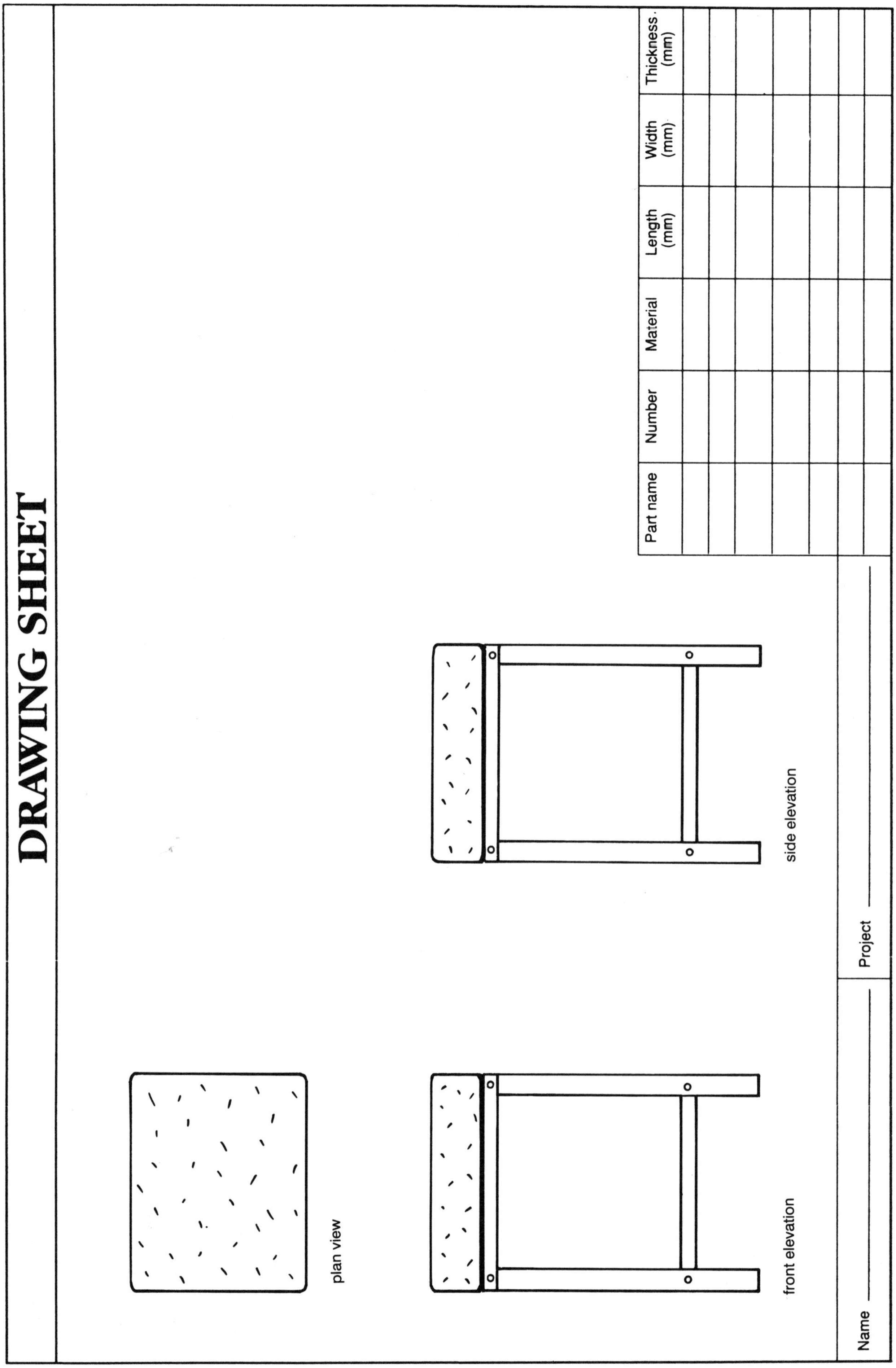

Fig. 6.14 An orthogonal drawing of a stool

STUDENT ACTIVITIES

1 Collect pictures of textile garments and objects that show good use of the following (two examples per area).
A. Line B. Colour combinations
C. Shape D. Pattern
2 Select a product you use regularly at home and answer the function analysis questions outlined in this chapter. Add at least two more questions of your own.
3 Think of two things you use every day that you find attractive or appealing and two that you don't. Explain why. See if you can redesign the unappealing object.
4 Write down as many uses as you can think of for a clothes peg.
5 Do a survey of the heights of people in the class. Measure them foot-knee, knee-waist, waist-shoulder. Present your findings graphically.
6 Briefly explain what would be in a Design Brief.
7 What do we mean by creating a balance in design?
8 What do we do in a Design Plan?
9 Ergonomics is the study of the relationship between people and their environment. True or False?
10 Aesthetics is to do with the function of a product. True or False?
11 Name the elements of Design.
12 Which is used to tint? Black or White?
13 Horizontal lines go up and down. True or False?
14 Complete a simple line drawing of a product. Photocopy or trace it three times on separate sheets, and complete the following.
a function analysis
an ergonomic analysis
an aesthetic analysis
These can be annotated drawings.

PRODUCTION: REALISATION AND EVALUATION

Once you have selected a final design, you need to make it. This is called the production or realisation. It may include analysing materials that we intend to use to ensure that we have selected the most appropriate ones. You will have discovered some things about materials testing in Chapter 5.

The final stage is evaluation. This is considered to be one of the most important stages, as it is through evaluating the product that we can determine the degree of success we have achieved. During this chapter we will briefly look at samples, and their production but the main emphasis will be on evaluation.

This chapter will focus on:

- prototypes, mock-ups and models
- production
- journal
- evaluation.

The last chapter outlined methods of testing and selecting materials and processes to be used. It may be necessary to make samples or models to test before making the actual product.

PROTOTYPES, MOCK-UPS AND MODELS

Prototypes, mock-ups and models are all methods of testing the materials, processes and equipment we wish to use. They are particularly important if you are using a new material, machine or process as you cannot be sure of the result until you test it.

A sample usually tests one section of your product. For instance, you may need to test a variety of seams on a new fabric you are interested in using. To find a result, you only need to complete each type of seam once—you don't need to complete the whole garment. Another instance where you will usually need to complete at least one sample is in decoration. You don't need to complete the whole decorative piece, but you must test it to ensure that the method suits the fabric.

A mock-up or model is usually a completed product but often using a different material. If the product is large, it may be a scale model of the product.

Mock-ups and models are made to see what the shape, size, proportion and overall look will be. In textiles, a mock-up of a garment may be made from calico or a cheap cotton fabric.

Once you are satisfied that you have completed all the tests you need you can make the final product.

Fig. 7.1 Hand painting on a shirt. Use a scrap of the same material for testing the paints.

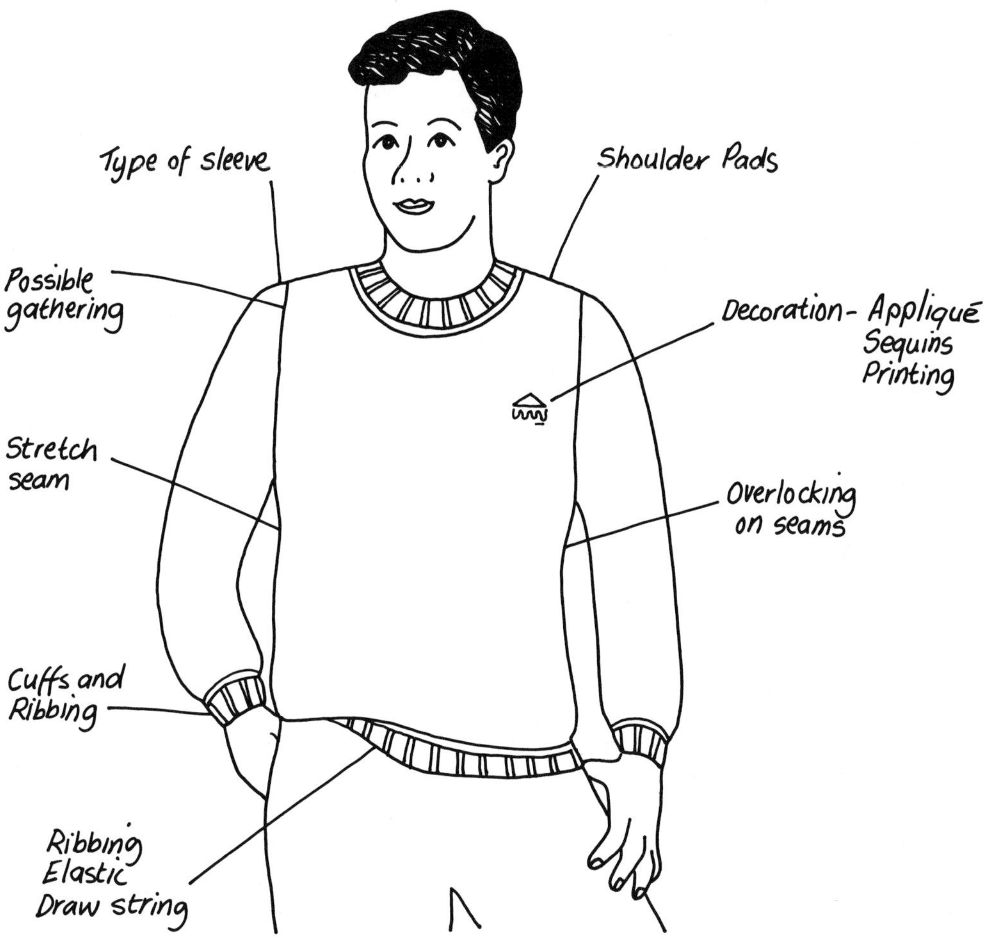

Fig. 7.2 Example of a windcheater and design considerations.

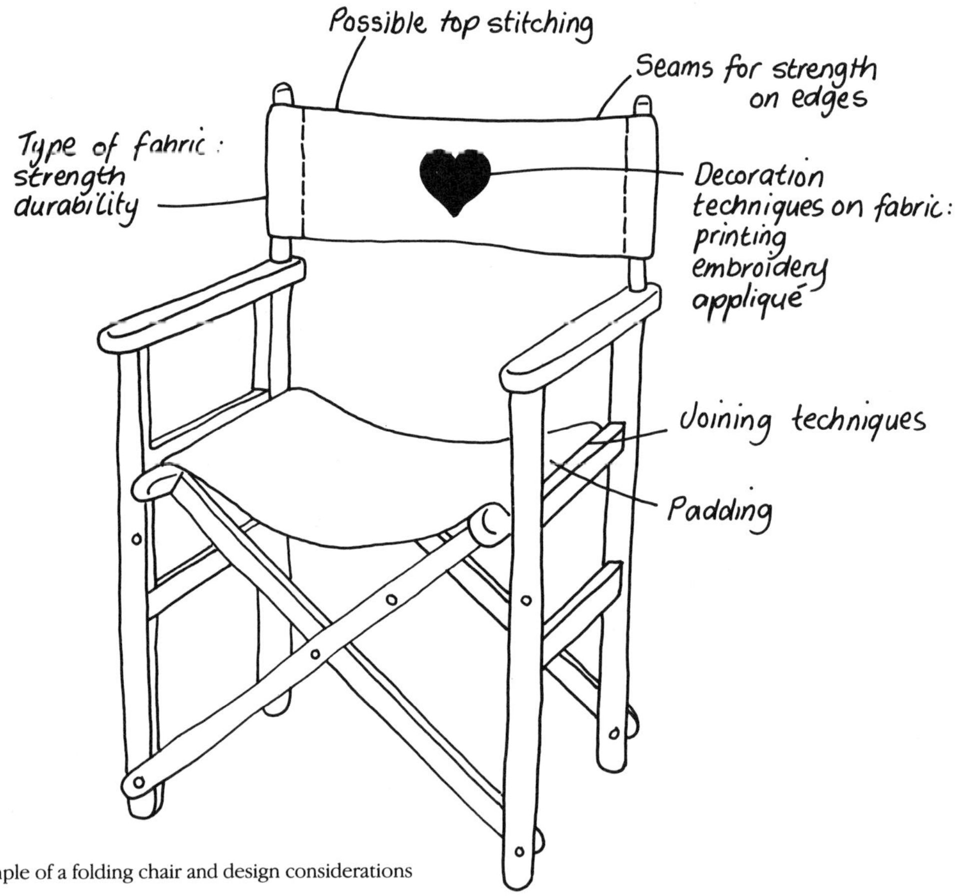

Fig. 7.3 Example of a folding chair and design considerations

Samples and testing are probably the areas we most like to avoid, as we are usually wanting to move on to making the final product, but be careful, spending a little extra time at this stage will save a lot of time, effort and frustration.

PRODUCTION

Production or realisation simply refers to the finished product or garment. Having completed all testing of materials and processes, making the final piece should be relatively straightforward, although it will be time consuming. It is important to allow enough time for this as you will find that you will run into difficulties if you try to rush the final production.

KEEPING A JOURNAL

A journal is a record of the tasks and processes you have followed throughout a project. Although this may seem simple, it is a vital part of any project. A journal should include a clear record of the work you have done from the Design Brief to Evaluation, with particular emphasis given to details of the processes followed in the production. It should also clearly record any alterations or problems and ways these were overcome (and there will always be many of these!).

There are a number of advantages in keeping a journal:

- It gives you a clear record of what you have done and any changes that you had to make.
- It can be referred to later for ideas on processes and other projects.
- It can be referred to by the 'client' to show them what you have done.
- Your teacher or assessor can gain some insight into your working pattern.

Fig. 7.4 A rushed finished product

JOURNAL / LOGBOOK ENTRY

NAME Petra Lozal (11B) DATE 16-3-92

PROJECT Tracksuit

Work done this week/session:

I have been very busy cutting my material from the patterns I made up last week. I needed to check all of the parts and make sure they were correct. I also enquired about how I am going to join the parts and Mrs Hailey is going to demonstrate this next lesson.

Alterations:

The only alteration I have made is a change of colour (to the school colours, I did this last week) I made this change as I have been picked in the school athletics team and would like to use the tracksuit when I'm competing for the school.

Plans for next week/session:

Mrs Hailey is going to demonstrate the way to sew the parts together. I have booked a machine for next lesson and hope to have part of the tracksuit put together then. I also need to purchase two zips for the legs as the school does not supply these.

Teacher's signature: Mrs. K. Hailey

Fig. 7.5 Example of a journal

During any design process you will find that you need to change some of your planned tasks as you evaluate your progress and discover different options may be better to follow. Therefore, there will be parts of your design plan which may not be completely accurate any more. Your journal is the place to record these changes. In all but a few design situations you will need to change some of your initial plans. This should be seen as an essential part of working through problems and not as a mistake.

There are numerous ways in which you can set up a journal. It may simply be a diary in an exercise book or on foolscap paper, or it may be more structured. Figure 7.5 gives an example of how you might set up a journal.

EVALUATION

Although the main evaluation is done on completion of the product, you should also evaluate the steps you have taken to ensure that any necessary changes are made and that you are on target. It will mean that you end up with a more successful final product. Your journal is where you record these changes. Evaluating is a vital step in the design process as it our way of reflecting on the success of the product. In this section we will look at who evaluates, what to evaluate and how to evaluate.

WHO EVALUATES?

Firstly, it is vital that you, the designer, evaluate the product. If the product is not for yourself then the intended user should evaluate it. Often you are asked by a client to design a product to their specifications or needs, so they should also evaluate it. Therefore there can be three different people or groups of people evaluating the product.

An example of different situations may be:

- You are designing a cushion for your desk chair at home. Therefore you evaluate it (as the designer).
- You are designing a cushion for a chair in your lounge room at home. You evaluate it (as the designer), and your family evaluate it (as the users).
- Your teacher has given you a design brief to design a cushion for school chairs. You evaluate it (as the designer), the students evaluate it (as the users), and your teacher evaluates it (as the client).

HOW TO EVALUATE

Usually the most efficient method of evaluating is by asking specific questions. A good starting point is to check your Design Brief and see what the considerations and limitations were. Once you have worked out your questions, you can complete your evaluation. This can be done through a survey, questionnaire or interview. Usually, as part of the design process, you will be asked to come up with your evaluation questions as part of your Design Plan. Below is one of the Design Briefs used in Chapter 6. See how you may go about evaluating the product.

DESIGN BRIEF

Students at your school are forever complaining about hard chairs. Your brief is to design something that will make them more comfortable.

Considerations/Limitations: cost, comfort, wear and tear, suitability of the chairs.

Designer Questions: Are they easy to make? Are you happy with the cost and weight of the fabric? Will they wear well? Do you like the final look of the chairs?

User Questions: Are the seats now more comfortable? Do they seem to wear well? Do they fit the chair? Do they look attractive?

Client Questions: Are you happy with the cost? Have you had any complaints about the comfort? Do they suit the chair? Do they fit the chair well? Do they look as though they will wear well?

Fig. 7.6 Who evaluates?

A CASE STUDY

PRODUCTION OF A WOOLLEN JACKET

By Julie Sultana

Many designers and dressmakers allow a lot of time for planning and consulting when working for a client. The structure of the Technology course I was involved in allowed much room for this, which made the process I went through to complete the garment take a more professional approach.

One of the three main areas of the process was the planning. This involved a lot of client consultation making sure the client's needs for the garment were going to be covered. From this, I developed a series of annotated designs accompanied by fabric samples, from which the client chose a design. I worked towards the design by planning a step by step guide to making that garment and a cost sheet for the client, which were collated with the designs in a work plan.

The second biggest process was the construction of the garment. This is where the work plan became a guide to refer to, helping me keep up with my initial schedule.

The most important element when working for a client is the amount of consultation that goes on. I was back and forth to my client showing my progress with the garment. I would allow him to try it on and received feedback from him, allowing room for alterations and changes to procedures. This is perhaps what made the whole process so good—the way we acted on a professional level. I gained the trust of the client and was able to work at a level which covered all the required needs of the garment.

The whole process towards finally completing the garment is very time consuming, but no step can be skipped. It was an individual learning experience, giving me insight towards the many hours of planning, preparing and making of a garment for a client or industry. This was a good chance for individual skills to be learnt and put into practice. I found that completing this design process was a worthwhile experience from which I gained a great deal in skills and knowledge. I feel pleased with my final garment and the experiences I have had. I am also glad that my client is satisfied with the jacket.

STUDENT ACTIVITIES

1 Below is one of the Design Briefs from Chapter 6. You are to design an outfit for your 20 year old sister who is attending an afternoon wedding in Melbourne in June.
Considerations/Limitations
Climate, time of day, cost, wearability, age
Possible Criteria for Evaluation
Designer: Are you happy with the final result? Do you feel it suits the client? Is it flexible or able to be worn again? How much did it cost? Does it seem reasonable? Would you alter any of the processes used? Did the processes suit the fabric? Would you alter any of them?
User/Client Questions: Are you pleased with the final garment? Do you think the fabric suits the design? Is the garment suitable for the occasion and climate? Do you think you will wear it again? Are you happy with the cost of it? Are you happy with the style?
Possible Design Plan
a Write a Design Brief (consult client)
b Write a Design Plan and Criteria for Evaluation
c Brainstorm/investigate
d Complete a series of trial solutions
e Select a preferred option with client
f Test materials, make samples
g Begin making garment
h Complete decoration and finishing
i Evaluate with client.
Use this brief to complete the trial solutions and select the preferred option. When completing the drawing of the preferred option, make sure you explain the various design elements of the garment.

2 You have been given the following Design Brief: Your school has many Bernina sewing machines which do not have covers. Design a cover to protect them from dust.
Considerations/Limitations:
Must be soft and flexible, need to fit standard Bernina sewing machine, cheap, easy to produce, easy to replace, with pockets to hold cord and foot.
Use the brief to complete the following:
- A Design Plan
- Criteria for Evaluation
- At least six trial solutions
- Choose a preferred option

3 Name the different people who may evaluate a product. Explain why.

4 Why is it important to complete samples and models for a product?

5 What must you consider when you are evaluating a product?

6 Why would you make a model, mock-up or prototype?

7 What is the production?

8 What is a journal? List two reasons why you would use a journal.

9 When do you evaluate?

INVESTIGATIONS

Investigating areas of technology gives us a wider perspective on what is happening around us. There are many interesting areas of technology to investigate. This chapter will help you to discover some of them and think about how to investigate them, and how to present your information in a clear and interesting way. We will also have a brief look at the area of marketing. Once a designer has designed a product, it needs to be sold. Marketing is the way in which this is done.

This chapter will focus on:

- **some legal aspects of textiles**
- **investigations: the research process; what to investigate, how to investigate and where to investigate and how to present your information**
- **methods of marketing.**

Throughout this century we have seen major technological developments which have had a huge effect on textiles and their uses. There are many important social factors that must be taken into consideration when we analyse the use of technology. In this chapter we will briefly discuss some pertinent issues.

The Australian textiles, clothing and footwear industries have undergone quite radical reforms and become increasingly politicised; for example, the use of tariffs to protect the Australian manufacturer from cheap offshore imports. There is no doubt we will continue to see more changes as computers and other electronic forms of equipment are developed and implemented into industry and the workforce. Manufacturing processes, marketing strategies, fibre development and environmental concerns will also continue to influence these changes. The primary economic principle which underlies any decision a manufacturer makes is: will it meet the consumer's demand and can it be manufactured, marketed and sold at a profit? Following such broad decisions it is necessary to consider the effect of new technologies on employment. The following questions should provide a basis for further discussion.

- How has the nature of the task (job) been affected by technological advances?

- Has unemployment occurred as a result of new methods and equipment?
- If jobs are lost, can people afford to purchase products?
- Do we need to become multi-skilled, less skilled or more specialised to perform the tasks involved?
- What new jobs have been created as a result of advances in technology and research relating to textiles?
- What is the role of tariffs in the Australian textiles, clothing and footwear industries?
- What assistance has the Government given to the development of more efficient, competitive and internationally oriented textiles, clothing and footwear industries?
- Australia has substantial resources, particularly in wool, cotton and hides, but we fail to transform any great proportion of these into finished products. Should we aim to improve our efficiency in these industries?

All of these issues are currently evident in the Australian textiles market and would provide an interesting debate in relation to various market sectors and industries. Consider in particular the implications they have for Australian employers, employees and consumers.

LEGAL ASPECTS OF THE TEXTILES INDUSTRY

In an industrialised society such as Australia, there is a need to regulate various aspects of manufacturing processes relating to materials and products. Consumer protection laws ensure the quality, performance and safety of the various products we purchase. You should make yourself familiar with the various Australian standards and legislation that cover such areas as labelling (which includes fibre content, care instructions, product safety, country of origin and sizing), manufacturing standards, health and safety and advertising.

There are also numerous agencies which deal with consumer complaints and assist in resolving disputes between consumers and manufacturers. You should investigate the role of the following Government and non-Government agencies:

- The Consumer Claims Tribunal
- The Australian Trade Practices Commission
- The Australian Consumers' Association
- The Consumer Affairs Council
- The Standards Association of Australia

CONSUMER DEMAND

As you are well aware, we love to make choices regarding the clothes we wear. Variety, therefore, is highly desirable. As a result of fast changing fashions we have seen the emergence of low cost garments which are readily replaced and therefore avert boredom. A range of garments are also needed for the changes in our climate throughout the year. Disposable items have gained popularity, particularly those items we wear for a season and then discard in place of a new fashion. Analyse your own wardrobe in terms of garments that you have purchased cheaply to be worn for a season and those clothes that are intended to last a lifetime or many seasons and deemed more 'Classic' styles.

MARKETING

For a design to be commercially successful, marketing strategies need to be employed to encourage consumers to buy the product. Many different marketing strategies have been developed. In this section we highlight some of them.

Of the various aspects of marketing, perhaps the packaging and advertising areas are the most relevant to us. We will look at them in more detail.

ADVERTISING

Advertising is how we attract consumers' attention and interest them in a product. It is the main link between the manufacturer and the consumer, but it is selective. The manufacturer projects the image they choose. There are many methods of advertising.

Different methods or media are employed to attract a different audience. The theme will also differ depending on who the advertisement is aiming to attract. There are a number of different forms an advertisement can take.

Fig. 8.1 Some examples of advertising

Fig. 8.2 Sex appeal in advertising

Fig. 8.3 Using status in advertising

PACKAGING

Packages take many different forms. These will vary depending on the product being sold and the image the seller is wishing to project.

Packaging can have three purposes:

1 To protect the product
2 To provide information about the product
3 To sell the product

Fig. 8.4 Packaging can be done in many different ways

Fig. 8.5 Packaging can protect the product

Fig. 8.6 Packaging can provide information about a product

YOU THE CONSUMER

We all buy products. What we buy and how much we buy is affected by the marketing strategies used. Consider those times when you have purchased something you initially had no intention of buying. It is likely that the marketing of that product was one of the reasons why you purchased it. We need to be selective in our buying, and as designers we should be considering the many facets of the product before we buy it. Some of the areas we should consider are:

- safety
- defects
- quality
- suitability
- features
- price
- cleaning
- materials used
- care required

It is difficult to consider this checklist each time you go shopping, but at least try to be selective about the quality of what you are buying.

At the end of this section are some helpful activities for finding out more about investigating and marketing.

Fig. 8.7 Packaging can be designed to attract the consumer's attention

INVESTIGATIONS

Investigating and research can be fun. Like most things it is crucial to keep yourself organised and follow through the process in a logical order.

Through investigating you will broaden your knowledge of issues arising from design and development. As you begin to research some of the many interesting topics you will discover how design and technology influence our lives in many different ways.

HOW TO RESEARCH

Research is a method of discovering. The key to research is to begin with a broad focus and work back to the details of your topic. The most obvious way of gaining information is by asking questions.

In a similar way to the design process, research is a series of linked activities. If you follow these simple steps you will be surprised how interesting and rewarding research can be. The essential first steps in a research investigation are:

Introduction

- Clarify the topic: what are you investigating? How is it related to design and technology? What do you already know about this topic? Clarify the intent of the investigation

Focus the questions

- Are you doing an interview or survey? Ensure that you formulate clear unbiased questions. Are you writing to companies for information? What specific information do you need? Where will you get it?

How will you present your information?

- Written report. Class demonstration. Pictorial/Graphic presentation. Oral delivery. Role-play/Drama activity. See our suggestions for more ideas.

Collection of data and information

This will help you gain background knowledge of the topic.

- Reading, viewing, listening, questioning, comparing. Identify information relevant to your purpose.
- Read suggested reference material. Investigate primary sources of information (e.g. interview and survey people, manufacturers, producers).
- Use secondary sources of information (e.g. newspapers, articles, reference books, journals).

Analysis and interpretation of data

- Analyse the data and relate the findings to the purpose of the inquiry. For example, what are the social, economic and environmental implications of the design. You must relate the information to your initial questions and explore related issues.
  Analyse
  Compare
  Contrast
  Define
  Describe
  Discuss
  Evaluate
  Examine
  Illustrate
  Summarise
- Check that you only select information relevant to the topic—don't get sidetracked.
- Have you got everything you need? Do you need to look further?

Conclusions

- Summarise your main findings. Do not introduce new information in your conclusion. The conclusions should draw all the investigation together. The reader will remember the conclusion, so make sure it has impact.

Recommendations

- Can you make any recommendations from the information you have gathered?

References

- A list of all the references you used must be attached to your work, and ensure you footnote where applicable. Ask your teacher for specific instructions on bibliographic conventions.

Appendixes

- Attach an appendix of all primary data collected. This may include letters sent, copies of surveys, information sent to you and any other relevant information.

HOW TO PRESENT YOUR INFORMATION

There are numerous forms of presenting work. Some will be more appropriate than others to different topics.

The four main methods of presenting information are:

- written
- graphic
- oral
- audio-visual

And, of course, you can combine them.

It is crucial when introducing an investigation, regardless of the method in which it is presented, to have a clear outline of its content. All investigations require a concise introduction of the topic and the main areas you will explore. From there you can divide it into sections exploring different aspects. Ensure that you sum up your findings with a concise concluding statement.

WHAT TO RESEARCH

The following are some suggested topics for investigation. These include throw-away products, depletion of natural resources and conservation groups.

WHY SELECT THESE AREAS?

One of the major consequences of developments in technology has been the environmental implications of the increasing use of throw-away products. Despite the convenience of throw-away products, society is beginning to recognise the problems that they can create. As designers, we must take these environmental issues into account.

Where unnatural or synthetic materials are used, the issue of recycling and pollution arises. Rivers, oceans and the air are being polluted as a consequence of industrialisation. Pollution can include wastes from manufacture, noise and products with short lives.

When natural resources are used in the manufacturing and processing of new products, either as energy (oil and gas used in manufacture) or for raw materials (wood, metal etc.), the resources are being depleted more rapidly. The issues of pollution and recycling also occur. Until now, the issue of recycling has not been addressed.

Recently, we have seen the formation of a range of conservation and environmental groups, working on a range of initiatives, local to global. These have been a direct result of the increase of pollution and the depletion of natural resources. They aim to identify issues, increase public awareness, and change laws. These groups range in size, methods of action and their target audience.

In the following section we look in some detail at throw-away products, have a brief look at the depletion of natural resources and look at environmental groups.

THROW-AWAY PRODUCTS

WHAT IS A THROW-AWAY PRODUCT?

As a class, brainstorm all the different throw-away products you can. Use the following list to double check that your item is a genuine 'throw-away' product.

Fig. 8.8 Throw away products

- Does it have a relatively short life span?
- Is it made to be used a limited number of times, perhaps only once, and discarded?

Once you have selected your product, look at the list of questions below. You may need to use some or all of them.

- Describe the product you have chosen to investigate. Does it replace another product? Can you briefly describe its development?
- What are the reasons for its development as a throw-away product?
- What are the benefits and problems (advantages and disadvantages) of the new product compared with the one it displaced?

Consider the following questions for the original product and the new throw-away product. Use this information to compare the benefits and problems of each.

- What are the social benefits and problems (family, home, work, time, leisure, effect on people's lives)?
- What are the economic benefits and problems (money, industry, cost for the consumer, reduced production costs, creation of new jobs)?
- What are the environmental issues (waste, pollution, resources)?

Don't forget to list the resources you used to answer the questions.

DEPLETION OF NATURAL RESOURCES

Identify a resource and list its qualities and characteristics. Some questions you may ask are:

- What is it used for?
- Are there any alternatives? Are they available?
- What are the advantages and disadvantages, or benefits and problems, associated with the use of the non-renewable resource and its alternative? Consider pollution, the environment, the future, animals, cost, people.
- What are the consequences should this resource disappear?
- Do any groups exist encouraging public awareness of the issue?
- What can you do?

ENVIRONMENTAL GROUPS

Environmental groups are very interesting to research. They have many different ways of creating awareness amongst the public: through advertising, demonstrating, music etc. Can you think of others? Bands such as Midnight Oil are well known for their environmental influence.

For your research you will need to select an environmental group and ask the following questions.

- What have they achieved?
- What do they aim to achieve?

Fig. 8.9 Depletion of natural resources

- What is their basic structure and role?
- Are they a local, state, national or international organisation?
- Do they have specific interests?
- Where does their funding come from (from donations, membership etc.)?

NEW AND MODIFIED MATERIALS

Chapter 5 gave you a brief glance at a variety of new and modified materials used by Paddy Pallin. Perhaps if you look in the fabric or textile outlets you will discover more innovative fibre developments. The wool industry is also continually researching and developing new fibre treatments—one being Superwash wool, which has become very popular with school uniform manufacturers as it can be machine washed without special attention.

Space technology has also been responsible for new inventions in relation to fabrics. Think about the protective suits that astronauts wear and the research and vigorous testing that has produced these new products. There are many newly developed fabrics that would make exciting research topics. As a class, brainstorm some examples. These will provide you with a good starting point for your next research assignment.

WHERE TO FIND RESOURCES

You will need a combination of primary and secondary sources.

There are countless resources to use relating to your chosen topics. Consider:

- The manufacturer of the product.
- Other related industries and competitors.

Fig. 8.10 Environmental groups

- Interviews, over the phone, in person or through written questionnaires.
- Your own observations.
- Survey class members and the wider community. Make sure the audience is relevant to the topic. For example, your class members aren't likely to use disposable nappies but people at the local community baby health centre might.
- Newspapers and magazines.
- Conversations with teachers, family, and other students. This can be particularly useful for exploring the replaced product. Talk to your grandmother about pantyhose back in her day or during the war for example.

The resources must be relevant to the aim of the investigation. A detailed list must be included with your investigation.

PRIMARY AND SECONDARY SOURCES

Primary sources are personal interviews, surveys, questionnaires, photographs, drawings and pictures. Journals or theses that have a person's first hand account of issues are good primary sources.

Secondary sources include newspaper articles, encyclopedias and abstracts. They are summaries of information from primary sources.

STUDENT ACTIVITIES

1 Select a product and collect advertisements for it. Also collect a sample of its package. Discuss some of the marketing techniques used—for example; factual, emotive, sex appeal and status. Analyse these advertisements in terms of the cultural images they are communicating.
2 Develop a logo for a textiles company of your choice. It should clearly reflect the image of the company.
3 See what you can find out about the history of packaging. Select a product or type of product and research its package development over the last 20 years.
4 Collect four advertisements from a magazine or newspaper and answer the following questions.
What type of advertisement is it? (is it emotive, factual etc.?)
Who is it aimed at?
Is it convincing?
Is it appealing?
How is it trying to persuade you to buy the product?
5 Select a natural resource and find out what it is used for. List some alternative ways of achieving a similar result without using the natural resource.
6 Select an Environmental Group and research it according to the guidelines outlined.
7 Design a useful way to store plastic bags.
8 List the waste in your home. Suggest at least three ways some of it could be reused.
9 When investigating what are primary sources? Name three.
10 How should you finish off any investigation?
11 Name three methods of presenting an investigation.
12 Name two depleted natural resources.
13 Name two types of pollution.
14 What is a bibliography? Why should you use one?

MINI DICTIONARY

Aesthetics: the study of the nature of beauty.

Alteration: change or modification of an idea or project.

Androgynous: displaying qualities of both sexes.

Appliqué: decorative technique of applying one fabric to another and stitching it in place.

AWC: Australian Wool Corporation.

Batik: Javanese method of painting fabric with wax and then dyeing it.

Beading: stitching small balls and sequins onto fabric for ornamental effects.

Bias: the diagonal grain of a fabric.

Boll: seedpod of the cotton shrub.

Carbonising: process of applying acid, heat and pressure to wool to remove impurities.

Carding: process of 'brushing' fibres to untangle them and prepare them for spinning.

Cellulose: constituent of the cell walls of plants. Used to make rayon and some plastics.

Conservation: preservation of the environment and natural resources.

Cording: technique used to highlight seams or the application of cord to lace for added texture and appeal.

Cotton: a small shrub. The white downy fibrous material covering the seeds is used to make cotton fibre.

Criteria: principles and standards used for assessing a product or process.

CSIRO: Commonwealth Scientific and Industrial Research Organisation.

Darts: give shape to a garment and remove fabric from where it is not needed (e.g. waist or bust).

Decoration: any number of techniques applied to fabric of garments for ornamental purposes.

Design Brief: short, clear statement describing a need or problem.

Design Plan: sequence of events necessary to complete a project.

Design: the process of finding the best option or solution in a particular set of circumstances.

Dyeing: process of colouring fabrics.

Elastic: able to return to its original shape after being stretched.

Ergonomics: study of the relationship between people and their working environment.

Evaluation: method of reflecting on the process and finished product to assess its success.

Fabric: the main form of textiles. Fabrics are made from fibres, which are processed and then constructed to form fabric. Construction techniques are usually weaving or knitting.

Fibre: a single strand of hair from a sheep's back or a single strand from a cotton boll. Synthetic fibres are sometimes referred to as filaments.

Filament: see fibre

Fray: when the edge of a fabric becomes unravelled.

Function: purpose for which a product was designed.

Grading: technique of trimming fabrics to different widths to eliminate bulk.

Jacquard Weave: the most complicated weave. A design is produced in self or many colours on the surface of the cloth.

Journal: record of the tasks and processes implemented throughout the project.

Knitting: uses one thread or yarn to produce an elastic, stretchable fabric. See also Warp Knitting and Weft Knitting.

Line: basic design element used to create style, shape, pattern and form.

Loom: machine used to weave fabric.

Marketing: promoting and selling ideas and products.

Mohair: hair from the Angora goat.

Natural: fibre derived from a natural source using only physical processes, such as carding, spinning etc. Fabric made from such fibres.

Notion: dressmaking extras (zips, buttons, interfacing, trims etc.).

Nylon: synthetic plastic material used to make clothing.

Orthogonal: detailed drawings to show different angles or views of an object.

Overlocker: machine which simultaneously casts a stitch over the edge of fabric and cuts to provide a professional edge to fabric and prevent it from fraying.

Packaging: parcel, box, or wrapping in which goods are stored and/or sold.

Pattern: combination of different lines and/or shapes. Patterns can be regular or irregular.

Plain Weave: the simplest weave. The weft thread goes over and under one wrap thread.

Pollution: contamination of the environment.

Polyester: synthetic resin/fibre used to make fabric.

Polymerisation: the process of joining together molecules to form polymers.

Polymers: compound formed by a combination of a number of identical molecules of a similar substance.

Preferred Solution/Option: most appealing or suitable idea chosen to solve a design brief.

Printing: method of applying patterns to the surface of fabrics.

Production: the process of making a project or product.

Rayon: the first synthetic fibre. Rayon is made from cellulose.

Recycle: convert waste into material that can be used again.

Reference: source of information.

Rendered: presentation illustration to portray ideas.

Research: discovering facts by study or critical investigation.

Right Side: the side of a fabric which will be seen in the final product.

Satin Weave: an irregular weave which leaves long 'floating' warp threads on the right side of the fabric. These floating threads catch the light and give satin its characteristic shiny appearance.

Scouring: process of removing dirt and grease from wool.

Seam: stitching line joining two or more pieces of fabric.

Selvedge: the edge of woven cloth finished to ensure it doesn't unravel or fray (self edge).

Spinning: the twisting together of many fibres or filaments to form a continuous thread or yarn.

Stretch: to extend, enlarge or distend. For a material to stretch implies that it will not return to its original size or shape.

Synthetic: fibre derived from either chemical or natural (usually cellulose) sources using chemical processes. Fabric made from such fibres.

Tanning: process of converting hides and skins into leather.

Textiles: the word originally applied to woven or plaited materials but now also covers knitted goods, leather, felt etc.

Texture: character or 'feel' of fabric resulting from the way it is constructed.

Thread: fibres or filaments spun together to form a long, thin cord (e.g. sewing cotton).

Tie-dyeing: dyeing technique used to produce unusual patterns.

Twill Weave: produces diagonal lines on the face of the cloth. The weft thread goes over two and under one warp thread.

Vilene: a non-woven fabric used to support and add body to fabrics (used in collars, cuffs, waistbands etc.).

Warp: one of the two threads used in weaving. Warp threads run parallel to the selvedge and run the full length of the material.

Warp Knitting: a row of loops are formed vertically using a number of threads, each controlled by its own needle.

Waste: materials that have no use which are left over or produced as a by-product of a process.

Weaving: the technique of making fabric by interlacing two sets of threads (warp and weft) at right angles.

Weft: one of the two threads used in weaving. Weft threads run at right angles to the selvedge and are usually weaker threads than the warp.

Weft Knitting: a single yarn runs horizontally making a row of loops into which the following row of loops are knitted.

Wool: fine soft crimped hair which forms the fleecy coat on sheep. Fibre and yarn made from that hair.

Woollen: shorter woollen fibres more commonly used for blankets and knitting yarns.

Worsted: long staple fibres of the same length tightly twisted to give a smooth surface to fabrics.

Wrong Side: the side of a fabric which will not be seen in the final product.

Yarn: thread made by twisting fibres together (e.g. knitting wool).

INDEX